Austin
Ambassador
Owners
Workshop
Manual

Peter G Strasman

Models covered
Austin Ambassador 1.7 L, HL; 1700 cc
Austin Ambassador 2.0 HL, HLS, Vanden Plas; 1994 cc

ISBN 0 85696 871 4

ABCD

Printed in England   *(871-10N2)*

Haynes Publishing Group
Sparkford Nr Yeovil
Somerset BA22 7JJ England

Haynes Publications, Inc
861 Lawrence Drive
Newbury Park
California 91320 USA

Acknowledgements

Thanks are due to BL Cars Limited for the supply of technical information and to Unipart and Lucas Electrical Limited for their assistance. The Champion Sparking Plug Company supplied the illustrations showing the various spark plug conditions. Duckhams Oils provided lubrication data, and Sykes-Pickavant Ltd provided some of the workshop tools. Special thanks are due to all those people at Sparkford who helped in the production of this manual.

About this manual

Its aim

The aim of this manual is to help you get the best value from your vehicle. It can do so in several ways. It can help you decide what work must be done (even should you choose to get it done by a garage), provide information on routine maintenance and servicing, and give a logical course of action and diagnosis when random faults occur. However, it is hoped that you will use the manual by tackling the work yourself. On simpler jobs it may even be quicker than booking the car into a garage and going there twice, to leave and collect it. Perhaps most important, a lot of money can be saved by avoiding the costs a garage must charge to cover its labour and overheads.

The manual has drawings and descriptions to show the function of the various components so that their layout can be understood. Then the tasks are described and photographed in a step-by-step sequence so that even a novice can do the work.

Its arrangement

The manual is divided into twelve Chapters, each covering a logical sub-division of the vehicle. The Chapters are each divided into Sections, numbered with single figures, eg 5; and the Sections into paragraphs (or sub-sections), with decimal numbers following on from the Section they are in, eg 5.1, 5.2, 5.3 etc.

It is freely illustrated, especially in those parts where there is a detailed sequence of operations to be carried out. There are two forms of illustration: figures and photographs. The figures are numbered in sequence with decimal numbers, according to their position in the Chapter – eg Fig. 6.4 is the fourth drawing/illustration in Chapter 6. Photographs carry the same number (either individually or in related groups) as the Section or sub-section to which they relate.

There is an alphabetical index at the back of the manual as well as a contents list at the front. Each Chapter is also preceded by its own individual contents list.

References to the 'left' or 'right' of the vehicle are in the sense of a person in the driver's seat facing forwards.

Unless otherwise stated, nuts and bolts are removed by turning anti-clockwise, and tightened by turning clockwise.

Vehicle manufacturers continually make changes to specifications and recommendations, and these, when notified, are incorporated into our manuals at the earliest opportunity.

Whilst every care is taken to ensure that the information in this manual is correct, no liability can be accepted by the authors or publishers for loss, damage or injury caused by any errors in, or omissions from, the information given.

Introduction to the Ambassador

The Ambassador is a much improved version of the Princess 2. It is a capacious five-door hatchback, surprisingly economical for such a large car.

The most impressive feature of the Ambassador is its ride comfort. The level of trim and equipment is very high.

The car is economical to run and the home mechanic will find it conventional to service and repair with plenty of room to work and excellent under-bonnet access.

Contents

Austin Ambassador HL

Austin Ambassador HLS

General dimensions, weights and capacities

Dimensions

Overall length	14.93 ft (4.55 m)
Overall width (two door mirrors fitted)	5.77 ft (1.76 m)
Overall height (kerb weight)	4.59 ft (1.40 m)
Ground clearance (kerb weight)	6.7 in (170.2 mm)
Wheelbase	8.76 ft (2.67 m)
Track (front)	4.86 ft (1.48 m)
Track (rear)	4.79 ft (1.46 m)
Luggage compartment capacity	12.8 ft^3 (0.36 m^3)

Weights (kerb with full fuel tank)

	Manual	Automatic
1.7L	2639 lb (1196 kg)	2655 lb (1204 kg)
1.7 HL	2653 lb (1203 kg)	2670 lb (1211 kg)
2.0 HL	2653 lb (1203 kg)	2670 lb (1211 kg)
2.0 HLS	2710 lb (1229 kg)	2728 lb (1237 kg)
Vanden Plas	2741 lb (1243 kg)	2761 lb (1252 kg)
Maximum towing weight (1 in 8 – 12% gradient in 1st speed gear)	2240 lb (1016 kg)	
Maximum roof rack load	110 lb (50 kg)	

Capacities

Fuel tank	16 gal (72.7 litre)
Cooling system	12.25 pt (6.95 litre)
Engine/manual transmission	10.25 pt (5.8 litre)
Engine (automatic transmission)	6.50 pt (3.7 litre)
Automatic transmission (from dry)	13.0 pt (7.4 litre)
Automatic transmission (drain and refill)	8.0 pt (4.5 litre)
Retained in torque converter	5.0 pt (2.8 litre)

Buying spare parts and vehicle identification numbers

Buying spare parts

Spare parts are available from many sources, for example: Austin-Rover garages, other garages and accessory shops, and motor factors. Our advice regarding spare parts is as follows:

Officially appointed Austin-Rover garages: This is the best source of parts which are peculiar to your car and otherwise not generally available (eg complete cylinder heads, internal gearbox components, badges, interior trim etc). It is also the only place at which you should buy parts if your car is still under warranty; non-Austin-Rover components may invalidate the warranty. To be sure of obtaining the correct parts it will always be necessary to give the storeman your car's engine and chassis number, and if possible to take the old part along for positive identification. Remember that many parts are available on a factory exchange scheme — any parts returned should always be clean! It obviously makes good sense to go to the specialists on your car for this type of part as they are best equipped to supply you.

Other garages and accessory shops — These are often very good places to buy material and components needed for the maintenance of your car (eg oil filters, spark plugs, bulbs, fan belts, oils and grease, touch-up paint, filler paste etc). They also sell general accessories, usually have convenient opening hours, charge lower prices and can often be found not far from home.

Motor factors — Good factors will stock all of the more important components which wear out relatively quickly (eg clutch components, pistons, valves, exhaust systems, brake cylinders/pipe/hoses/seals/shoes and pads etc). Motor factors will often provide new or reconditioned components on a part exchange basis — this can save a considerable amount of money.

Vehicle identification numbers

Modifications are a continuing and unpublicised process in vehicle manufacture quite apart from their major model changes. Spare parts manuals and lists are compiled on a numerical basis, the individual vehicle number being essential for correct identification of the component required.

The vehicle number may be obtained from the identification plate within the engine compartment and the *engine number* is stamped on a metal plate on the cylinder block. Both numbers are also shown in the vehicle registration document. The body number is stamped on a plate fixed to the left-hand end of the bonnet lock platform. If the car is equipped with automatic transmission, the transmission number is located on a plate on the torque converter above the starter motor.

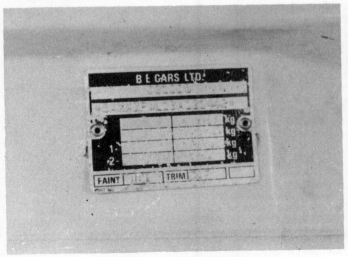

Vehicle identification number

Engine number

Tools and working facilities

Introduction

A selection of good tools is a fundamental requirement for anyone contemplating the maintenance and repair of a motor vehicle. For the owner who does not possess any, their purchase will prove a considerable expense, offsetting some of the savings made by doing-it-yourself. However, provided that the tools purchased meet the relevant national safety standards and are of good quality, they will last for many years and prove an extremely worthwhile investment.

To help the average owner to decide which tools are needed to carry out the various tasks detailed in this manual, we have compiled three lists of tools under the following headings: *Maintenance and minor repair, Repair and overhaul,* and *Special.* The newcomer to practical mechanics should start off with the *Maintenance and minor repair* tool kit and confine himself to the simpler jobs around the vehicle. Then, as his confidence and experience grow, he can undertake more difficult tasks, buying extra tools as, and when, they are needed. In this way, a *Maintenance and minor repair* tool kit can be built-up into a *Repair and overhaul* tool kit over a considerable period of time without any major cash outlays. The experienced do-it-yourselfer will have a tool kit good enough for most repair and overhaul procedures and will add tools from the *Special* category when he feels the expense is justified by the amount of use to which these tools will be put.

It is obviously not possible to cover the subject of tools fully here. For those who wish to learn more about tools and their use there is a book entitled *How to Choose and Use Car Tools* available from the publishers of this manual.

Maintenance and minor repair tool kit

The tools given in this list should be considered as a minimum requirement if routine maintenance, servicing and minor repair operations are to be undertaken. We recommend the purchase of combination spanners (ring one end, open-ended the other); although more expensive than open-ended ones, they do give the advantages of both types of spanner.

Combination spanners - 10, 11, 12, 13, 14 & 17 mm
Adjustable spanner - 9 inch
Spark plug spanner (with rubber insert)
Spark plug gap adjustment tool
Set of feeler gauges
Brake bleed nipple spanner
Screwdriver - 4 in long x $\frac{1}{4}$ in dia (flat blade)
Screwdriver - 4 in long x $\frac{1}{4}$ in dia (cross blade)
Combination pliers - 6 inch
Hacksaw (junior)
Tyre pump
Tyre pressure gauge
Grease gun
Oil can
Fine emery cloth (1 sheet)
Wire brush (small)
Funnel (medium size)

Repair and overhaul tool kit

These tools are virtually essential for anyone undertaking any major repairs to a motor vehicle, and are additional to those given in the *Maintenance and minor repair* list. Included in this list is a comprehensive set of sockets. Although these are expensive they will be found invaluable as they are so versatile - particularly if various drives are included in the set. We recommend the $\frac{1}{2}$ in square-drive type, as this can be used with most proprietary torque wrenches. If you

cannot afford a socket set, even bought piecemeal, then inexpensive tubular box spanners are a useful alternative.

The tools in this list will occasionally need to be supplemented by tools from the *Special* list.

Sockets (or box spanners) to cover range in previous list
Reversible ratchet drive (for use with sockets)
Extension piece, 10 inch (for use with sockets)
Universal joint (for use with sockets)
Torque wrench (for use with sockets)
'Mole' wrench - 8 inch
Ball pein hammer
Soft-faced hammer, plastic or rubber
Screwdriver - 6 in long x $\frac{5}{16}$ in dia (flat blade)
Screwdriver - 2 in long x $\frac{5}{16}$ in square (flat blade)
Screwdriver - 1$\frac{1}{2}$ in long x $\frac{1}{4}$ in dia (cross blade)
Screwdriver - 3 in long x $\frac{1}{8}$ in dia (electricians)
Pliers - electricians side cutters
Pliers - needle nosed
Pliers - circlip (internal and external)
Cold chisel - $\frac{1}{2}$ inch
Scriber
Scraper
Centre punch
Pin punch
Hacksaw
Valve grinding tool
Steel rule/straight-edge
Allen keys
Selection of files
Wire brush (large)
Axle-stands
Jack (strong scissor or hydraulic type)

Special tools

The tools in this list are those which are not used regularly, are expensive to buy, or which need to be used in accordance with their manufacturers' instructions. Unless relatively difficult mechanical jobs are undertaken frequently, it will not be economic to buy many of these tools. Where this is the case, you could consider clubbing together with friends (or joining a motorists' club) to make a joint purchase, or borrowing the tools against a deposit from a local garage or tool hire specialist.

The following list contains only those tools and instruments freely available to the public, and not those special tools produced by the vehicle manufacturer specifically for its dealer network. You will find occasional references to these manufacturers' special tools in the text of this manual. Generally, an alternative method of doing the job without the vehicle manufacturers' special tool is given. However, sometimes, there is no alternative to using them. Where this is the case and the relevant tool cannot be bought or borrowed, you will have to entrust the work to a franchised garage.

Valve spring compressor
Piston ring compressor
Balljoint separator
Universal hub/bearing puller
Impact screwdriver
Micrometer and/or vernier gauge
Dial gauge
Stroboscopic timing light
Dwell angle meter/tachometer

Universal electrical multi-meter
Cylinder compression gauge
Lifting tackle (photo)
Trolley jack
Light with extension lead

Buying tools

For practically all tools, a tool factor is the best source since he will have a very comprehensive range compared with the average garage or accessory shop. Having said that, accessory shops often offer excellent quality tools at discount prices, so it pays to shop around.

There are plenty of good tools around at reasonable prices, but always aim to purchase items which meet the relevant national safety standards. If in doubt, ask the proprietor or manager of the shop for advice before making a purchase.

Care and maintenance of tools

Having purchased a reasonable tool kit, it is necessary to keep the tools in a clean serviceable condition. After use, always wipe off any dirt, grease and metal particles using a clean, dry cloth, before putting the tools away. Never leave them lying around after they have been used. A simple tool rack on the garage or workshop wall, for items such as screwdrivers and pliers is a good idea. Store all normal wrenches and sockets in a metal box. Any measuring instruments, gauges, meters, etc, must be carefully stored where they cannot be damaged or become rusty.

Take a little care when tools are used. Hammer heads inevitably become marked and screwdrivers lose the keen edge on their blades from time to time. A little timely attention with emery cloth or a file will soon restore items like this to a good serviceable finish.

Working facilities

Not to be forgotten when discussing tools, is the workshop itself. If anything more than routine maintenance is to be carried out, some form of suitable working area becomes essential.

It is appreciated that many an owner mechanic is forced by circumstances to remove an engine or similar item, without the benefit of a garage or workshop. Having done this, any repairs should always be done under the cover of a roof.

Wherever possible, any dismantling should be done on a clean, flat workbench or table at a suitable working height.

Any workbench needs a vice: one with a jaw opening of 4 in (100 mm) is suitable for most jobs. As mentioned previously, some clean dry storage space is also required for tools, as well as for lubricants, cleaning fluids, touch-up paints and so on, which become necessary.

Another item which may be required, and which has a much more general usage, is an electric drill with a chuck capacity of at least $\frac{5}{16}$ in (8 mm). This, together with a good range of twist drills, is virtually essential for fitting accessories such as mirrors and reversing lights.

Last, but not least, always keep a supply of old newspapers and clean, lint-free rags available, and try to keep any working area as clean as possible.

Spanner jaw gap comparison table

Jaw gap (in)	Spanner size
0.250	$\frac{1}{4}$ in AF
0.276	7 mm
0.313	$\frac{5}{16}$ in AF
0.315	8 mm
0.344	$\frac{11}{32}$ in AF; $\frac{1}{8}$ in Whitworth
0.354	9 mm
0.375	$\frac{3}{8}$ in AF
0.394	10 mm
0.433	11 mm
0.438	$\frac{7}{16}$ in AF
0.445	$\frac{3}{16}$ in Whitworth; $\frac{1}{4}$ in BSF
0.472	12 mm
0.500	$\frac{1}{2}$ in AF
0.512	13 mm
0.525	$\frac{1}{4}$ in Whitworth; $\frac{5}{16}$ in BSF
0.551	14 mm
0.563	$\frac{9}{16}$ in AF
0.591	15 mm
0.600	$\frac{5}{16}$ in Whitworth; $\frac{3}{8}$ in BSF
0.625	$\frac{5}{8}$ in AF
0.630	16 mm
0.669	17 mm
0.686	$\frac{11}{16}$ in AF
0.709	18 mm
0.710	$\frac{3}{8}$ in Whitworth; $\frac{7}{16}$ in BSF
0.748	19 mm
0.750	$\frac{3}{4}$ in AF
0.813	$\frac{13}{16}$ in AF
0.820	$\frac{7}{16}$ in Whitworth; $\frac{1}{2}$ in BSF
0.866	22 mm
0.875	$\frac{7}{8}$ in AF
0.920	$\frac{1}{2}$ in Whitworth; $\frac{9}{16}$ in BSF
0.938	$\frac{15}{16}$ in AF
0.945	24 mm
1.000	1 in AF
1.010	$\frac{9}{16}$ in Whitworth; $\frac{5}{8}$ in BSF
1.024	26 mm
1.063	$1\frac{1}{16}$ in AF; 27 mm
1.100	$\frac{5}{8}$ in Whitworth; $\frac{11}{16}$ in BSF
1.125	$1\frac{1}{8}$ in AF
1.181	30 mm
1.200	$\frac{11}{16}$ in Whitworth; $\frac{3}{4}$ in BSF
1.250	$1\frac{1}{4}$ in AF
1.260	32 mm
1.300	$\frac{3}{4}$ in Whitworth; $\frac{7}{8}$ in BSF
1.313	$1\frac{5}{16}$ in AF
1.390	$\frac{13}{16}$ in Whitworth; $\frac{15}{16}$ in BSF
1.417	36 mm
1.438	$1\frac{7}{16}$ in AF
1.480	$\frac{7}{8}$ in Whitworth; 1 in BSF
1.500	$1\frac{1}{2}$ in AF
1.575	40 mm; $\frac{15}{16}$ in Whitworth
1.614	41 mm
1.625	$1\frac{5}{8}$ in AF
1.670	1 in Whitworth; $1\frac{1}{8}$ in BSF
1.688	$1\frac{11}{16}$ in AF
1.811	46 mm
1.813	$1\frac{13}{16}$ in AF
1.860	$1\frac{1}{8}$ in Whitworth; $1\frac{1}{4}$ in BSF
1.875	$1\frac{7}{8}$ in AF
1.969	50 mm
2.000	2 in AF
2.050	$1\frac{1}{4}$ in Whitworth; $1\frac{3}{8}$ in BSF
2.165	55 mm
2.362	60 mm

General repair procedures

Whenever servicing, repair or overhaul work is carried out on the car or its components, it is necessary to observe the following procedures and instructions. This will assist in carrying out the operation efficiently and to a professional standard of workmanship.

Joint mating faces and gaskets

Where a gasket is used between the mating faces of two components, ensure that it is renewed on reassembly, and fit it dry unless otherwise stated in the repair procedure. Make sure that the mating faces are clean and dry with all traces of old gasket removed. When cleaning a joint face, use a tool which is not likely to score or damage the face, and remove any burrs or nicks with an oilstone or fine file.

Make sure that tapped holes are cleaned with a pipe cleaner, and keep them free of jointing compound if this is being used unless specifically instructed otherwise.

Ensure that all orifices, channels or pipes are clear and blow through them, preferably using compressed air.

Oil seals

Whenever an oil seal is removed from its working location, either individually or as part of an assembly, it should be renewed.

The very fine sealing lip of the seal is easily damaged and will not seal if the surface it contacts is not completely clean and free from scratches, nicks or grooves. If the original sealing surface of the component cannot be restored, the component should be renewed.

Protect the lips of the seal from any surface which may damage them in the course of fitting. Use tape or a conical sleeve where possible. Lubricate the seal lips with oil before fitting and, on dual lipped seals, fill the space between the lips with grease.

Unless otherwise stated, oil seals must be fitted with their sealing lips toward the lubricant to be sealed.

Use a tubular drift or block of wood of the appropriate size to install the seal and, if the seal housing is shouldered, drive the seal down to the shoulder. If the seal housing is unshouldered, the seal should be fitted with its face flush with the housing top face.

Screw threads and fastenings

Always ensure that a blind tapped hole is completely free from oil, grease, water or other fluid before installing the bolt or stud. Failure to do this could cause the housing to crack due to the hydraulic action of the bolt or stud as it is screwed in.

When tightening a castellated nut to accept a split pin, tighten the nut to the specified torque, where applicable, and then tighten further to the next split pin hole. Never slacken the nut to align a split pin hole unless stated in the repair procedure.

When checking or retightening a nut or bolt to a specified torque setting, slacken the nut or bolt by a quarter of a turn, and then retighten to the specified setting.

Locknuts, locktabs and washers

Any fastening which will rotate against a component or housing in the course of tightening should always have a washer between it and the relevant component or housing.

Spring or split washers should always be renewed when they are used to lock a critical component such as a big-end bearing retaining nut or bolt.

Locktabs which are folded over to retain a nut or bolt should always be renewed.

Self-locking nuts can be reused in non-critical areas, providing resistance can be felt when the locking portion passes over the bolt or stud thread.

Split pins must always be replaced with new ones of the correct size for the hole.

Special tools

Some repair procedures in this manual entail the use of special tools such as a press, two or three-legged pullers, spring compressors etc. Wherever possible, suitable readily available alternatives to the manufacturer's special tools are described, and are shown in use. In some instances, where no alternative is possible, it has been necessary to resort to the use of a manufacturer's tool and this has been done for reasons of safety as well as the efficient completion of the repair operation. Unless you are highly skilled and have a thorough understanding of the procedure described, never attempt to bypass the use of any special tool when the procedure described specifies its use. Not only is there a very great risk of personal injury, but expensive damage could be caused to the components involved.

Jacking, towing and hoisting

The jack supplied with the car should only be used to raise the car for changing the roadwheels. If it is used in maintenance or repair operations, then it must be supplemented with axle stands. Two jacking points are provided on each side of the car; always chock the roadwheels not being raised. When using a workshop jack, it must only be located as described, otherwise distortion of the body may occur.

To avoid repetition, the procedure for raising the vehicle in order to carry out work under it is not included before each relevant operation described in this Manual.

It is to be preferred and it is certainly recommended that the vehicle is positioned over an inspection pit or raised on a lift. Where these facilities are not available, use ramps or jack up the vehicle strictly in accordance with the following guide and once the vehicle is raised, supplement the jack with axle stands.

To jack one front wheel, locate the jack under the reinforcement member (2) (see illustration) to the rear of the suspension lower arm rear bearing. Do not jack on the rear bearing itself.

To jack both fronts wheels at the same time, a steel section or baulk of timber should be placed across the reinforcement members and a jack placed under the centre of the temporary support.

To jack one rear wheel, locate the jack under the rear suspension bracket (3).

Wheel changing jack

Tool kit wheel chock

Spare wheel and cover

Spare wheel retainer

Tool compartment lid screw

Tool compartment

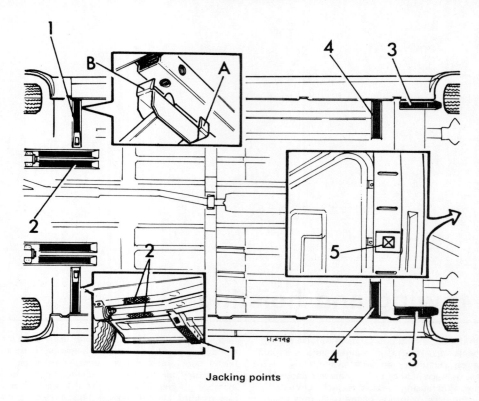

Jacking points

1	*Front jacking bracket (tool kit jack)*	4	*Rear jacking bracket (tool kit jack)*
2	*Front reinforcement member*	5	*Rear reinforcement bracket*
3	*Rear suspension bracket*	A and B	*Jack head positioning tabs*

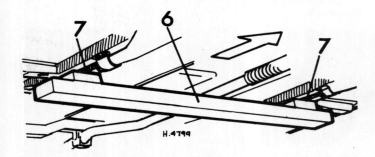

Cross support for lifting both front wheels at the same time

6   *Temporary supports*
7   *Spacers*

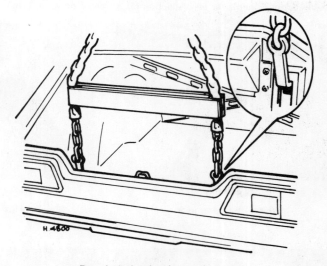

Rear hoisting hook attachment

To jack both rear wheels, locate the jack under the reinforcement plate (5) at the centre of the rear panel.

Whenever repairs or adjustments are being carried out under the car, always supplement the jacks with axle stands positioned under the wheel changing jacking brackets.

To hoist the rear of the car, hooks can be attached to the slots within the luggage boot and a spreader used.

To hoist the front of the car, hooks can be attached to the eyes within the engine compartment and a spreader used.

The car may be towed by another vehicle using the towing eyes located one each side of the front body member. A suspended tow must not be carried out using these eyes.

The towing eye attached to the rear jacking plate may be used to tow another car.

Where a caravan or trailer is to be towed, fit an approved towing kit and make sure that the total weight of the unit being towed does not exceed 2240 lbs (1016 kg).

If the car is equipped with automatic transmission, the car must not be towed by another vehicle unless an additional 3 Imp pints (1.7

litres) of automatic transmission fluid are added, the towing speed is restricted to below 30 mph (50 km/h) and the distance towed restricted to a maximum of 30 miles (50 km). When being towed, set the speed selector lever to 'N' and turn the ignition key to 'I'. If these conditions cannot be complied with, the front of the car must be raised and placed on a recovery trailer.

Before raising the car to remove a roadwheel, prise off the wheel nut covers or trim plate according to type. Slightly release the wheel nuts. Raise the car and unscrew and remove the nuts. When refitting, apply a little grease to the stud threads. Tighten the nuts in a clockwise sequence as shown.

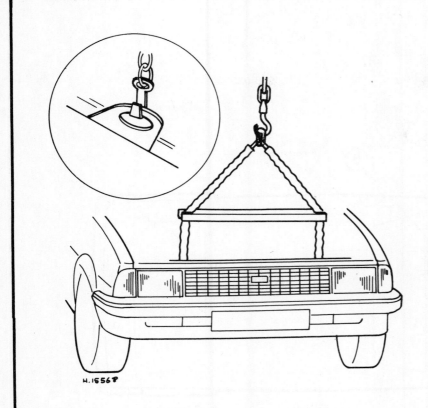

Front hoisting hook attachment

Front towing hook

Rear towing hook

Levering off wheel nut cap

Unscrewing a wheel nut

Removing a wheel trim

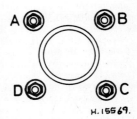

Roadwheel nut tightening sequence

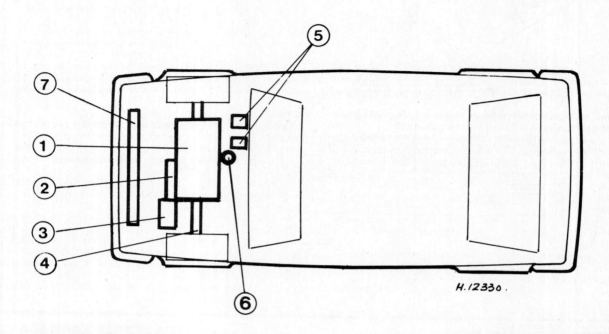

H./2330.

Recommended lubricants and fluids

Component or system	Lubricant type/specification	Duckhams recommendation
1 Engine (and manual transmission)*	Multigrade engine oil, viscosity SAE 10W/40, 15W/40 or 15W/50, to API SE/SF	Duckhams QXR, Hypergrade or 10W/40 Motor Oil
2 Automatic transmission	Automatic transmission fluid to M2C 33G	Duckhams Q-Matic
3 Power steering reservoir Up to VIN 122690	Multigrade engine oil, viscosity SAE 15W/50, to API SE/SF	Duckhams Hypergrade
From VIN 122691	Automatic transmission fluid to M2C 33G or M2C 33F	Duckhams Q-Matic
4 Steering rack	Multi-purpose lithium based grease to NLGI No 2	Duckhams LB 10
5 Brake and clutch fluid reservoirs	Hydraulic fluid to FMVSS DOT 3	Duckhams Universal Brake and Clutch Fluid
6 Carburettor damper	Multigrade engine oil, viscosity SAE 10W/40, 15W/40 or 15W/50, to API SE/SF	Duckhams QXR, Hypergrade or 10W/40 Motor Oil
7 Cooling system	Antifreeze to BS 3151, 3152 or 6580	·Duckhams Universal Antifreeze and Summer Coolant

*Note: Austin Rover specify a 10W/40 oil to meet warranty requirements for models produced after August 1983. Duckhams QXR and 10W/40 Motor Oil are available to meet these requirements

Safety first!

Professional motor mechanics are trained in safe working procedures. However enthusiastic you may be about getting on with the job in hand, do take the time to ensure that your safety is not put at risk. A moment's lack of attention can result in an accident, as can failure to observe certain elementary precautions.

There will always be new ways of having accidents, and the following points do not pretend to be a comprehensive list of all dangers; they are intended rather to make you aware of the risks and to encourage a safety-conscious approach to all work you carry out on your vehicle.

Essential DOs and DON'Ts

DON'T rely on a single jack when working underneath the vehicle. Always use reliable additional means of support, such as axle stands, securely placed under a part of the vehicle that you know will not give way.

DON'T attempt to loosen or tighten high-torque nuts (e.g. wheel hub nuts) while the vehicle is on a jack; it may be pulled off.

DON'T start the engine without first ascertaining that the transmission is in neutral (or 'Park' where applicable) and the parking brake applied.

DON'T suddenly remove the filler cap from a hot cooling system – cover it with a cloth and release the pressure gradually first, or you may get scalded by escaping coolant.

DON'T attempt to drain oil until you are sure it has cooled sufficiently to avoid scalding you.

DON'T grasp any part of the engine, exhaust or catalytic converter without first ascertaining that it is sufficiently cool to avoid burning you.

DON'T allow brake fluid or antifreeze to contact vehicle paintwork.

DON'T syphon toxic liquids such as fuel, brake fluid or antifreeze by mouth, or allow them to remain on your skin.

DON'T inhale dust – it may be injurious to health (see *Asbestos* below).

DON'T allow any spilt oil or grease to remain on the floor – wipe it up straight away, before someone slips on it.

DON'T use ill-fitting spanners or other tools which may slip and cause injury.

DON'T attempt to lift a heavy component which may be beyond your capability – get assistance.

DON'T rush to finish a job, or take unverified short cuts.

DON'T allow children or animals in or around an unattended vehicle.

DO wear eye protection when using power tools such as drill, sander, bench grinder etc, and when working under the vehicle.

DO use a barrier cream on your hands prior to undertaking dirty jobs – it will protect your skin from infection as well as making the dirt easier to remove afterwards; but make sure your hands aren't left slippery. Note that long-term contact with used engine oil can be a health hazard.

DO keep loose clothing (cuffs, tie etc) and long hair well out of the way of moving mechanical parts.

DO remove rings, wristwatch etc, before working on the vehicle – especially the electrical system.

DO ensure that any lifting tackle used has a safe working load rating adequate for the job.

DO keep your work area tidy – it is only too easy to fall over articles left lying around.

DO get someone to check periodically that all is well, when working alone on the vehicle.

DO carry out work in a logical sequence and check that everything is correctly assembled and tightened afterwards.

DO remember that your vehicle's safety affects that of yourself and others. If in doubt on any point, get specialist advice.

IF, in spite of following these precautions, you are unfortunate enough to injure yourself, seek medical attention as soon as possible.

Asbestos

Certain friction, insulating, sealing, and other products – such as brake linings, brake bands, clutch linings, torque converters, gaskets, etc – contain asbestos. *Extreme care must be taken to avoid inhalation of dust from such products since it is hazardous to health.* If in doubt, assume that they *do* contain asbestos.

Fire

Remember at all times that petrol (gasoline) is highly flammable. Never smoke, or have any kind of naked flame around, when working on the vehicle. But the risk does not end there – a spark caused by an electrical short-circuit, by two metal surfaces contacting each other, by careless use of tools, or even by static electricity built up in your body under certain conditions, can ignite petrol vapour, which in a confined space is highly explosive.

Always disconnect the battery earth (ground) terminal before working on any part of the fuel or electrical system, and never risk spilling fuel on to a hot engine or exhaust.

It is recommended that a fire extinguisher of a type suitable for fuel and electrical fires is kept handy in the garage or workplace at all times. Never try to extinguish a fuel or electrical fire with water.

Note: *Any reference to a 'torch' appearing in this manual should always be taken to mean a hand-held battery-operated electric lamp or flashlight. It does NOT mean a welding/gas torch or blowlamp.*

Fumes

Certain fumes are highly toxic and can quickly cause unconsciousness and even death if inhaled to any extent. Petrol (gasoline) vapour comes into this category, as do the vapours from certain solvents such as trichloroethylene. Any draining or pouring of such volatile fluids should be done in a well ventilated area.

When using cleaning fluids and solvents, read the instructions carefully. Never use materials from unmarked containers – they may give off poisonous vapours.

Never run the engine of a motor vehicle in an enclosed space such as a garage. Exhaust fumes contain carbon monoxide which is extremely poisonous; if you need to run the engine, always do so in the open air or at least have the rear of the vehicle outside the workplace.

If you are fortunate enough to have the use of an inspection pit, never drain or pour petrol, and never run the engine, while the vehicle is standing over it; the fumes, being heavier than air, will concentrate in the pit with possibly lethal results.

The battery

Never cause a spark, or allow a naked light, near the vehicle's battery. It will normally be giving off a certain amount of hydrogen gas, which is highly explosive.

Always disconnect the battery earth (ground) terminal before working on the fuel or electrical systems.

If possible, loosen the filler plugs or cover when charging the battery from an external source. Do not charge at an excessive rate or the battery may burst.

Take care when topping up and when carrying the battery. The acid electrolyte, even when diluted, is very corrosive and should not be allowed to contact the eyes or skin.

If you ever need to prepare electrolyte yourself, always add the acid slowly to the water, and never the other way round. Protect against splashes by wearing rubber gloves and goggles.

When jump starting a car using a booster battery, for negative earth (ground) vehicles, connect the jump leads in the following sequence: First connect one jump lead between the positive (+) terminals of the two batteries. Then connect the other jump lead first to the negative (–) terminal of the booster battery, and then to a good earthing (ground) point on the vehicle to be started, at least 18 in (45 cm) from the battery if possible. Ensure that hands and jump leads are clear of any moving parts, and that the two vehicles do not touch. Disconnect the leads in the reverse order.

Mains electricity and electrical equipment

When using an electric power tool, inspection light etc, always ensure that the appliance is correctly connected to its plug and that, where necessary, it is properly earthed (grounded). Do not use such appliances in damp conditions and, again, beware of creating a spark or applying excessive heat in the vicinity of fuel or fuel vapour. Also ensure that the appliances meet the relevant national safety standards.

Ignition HT voltage

A severe electric shock can result from touching certain parts of the ignition system, such as the HT leads, when the engine is running or being cranked, particularly if components are damp or the insulation is defective. Where an electronic ignition system is fitted, the HT voltage is much higher and could prove fatal.

Routine maintenance

Maintenance is essential for ensuring safety and desirable for the purpose of getting the best in terms of performance and economy from the car. Over the years the need for periodic lubrication – oiling, greasing and so on – has been drastically reduced if not totally eliminated. This has unfortunately tended to lead some owners to think that because no such action is required the items either no longer exist or will last for ever. This is a serious delusion. It follows therefore that the largest initial element of maintenance is visual examination. This may lead to repairs or renewals.

The following maintenance schedules are based upon the vehicle manufacturers recommendations, but certain checking intervals have been shortened where it is felt that servicing checks are so infrequent that a dangerous situation could occur (such as with brake pads and linings) if these items are not inspected regularly and frequently.

The frequency of engine oil and filter renewal has also been increased as a precaution against engine and transmission internal corrosion which can arise because of the formation of condensation and acidity as the result of short journey operation and piston ring blow-by of combustion gases.

Engine compartment

1 Brake master cylinder	5 Carburettor	8 Oil filler cap	11 Radiator
2 Vacuum servo unit	6 Battery	9 Distributor	12 Dipstick
3 Clutch master cylinder	7 Coolant expansion tank	10 Thermostat	13 Washer bottle
4 Air cleaner			

Underside of front end

1	Suspension lower arm	4 Steering gear	7 Transmission casing	9 Engine/transmission oil
2	Driveshaft	5 Exhaust downpipe	8 Power steering fluid	drain plug
3	Engine mounting	6 Gearchange rod	lines	

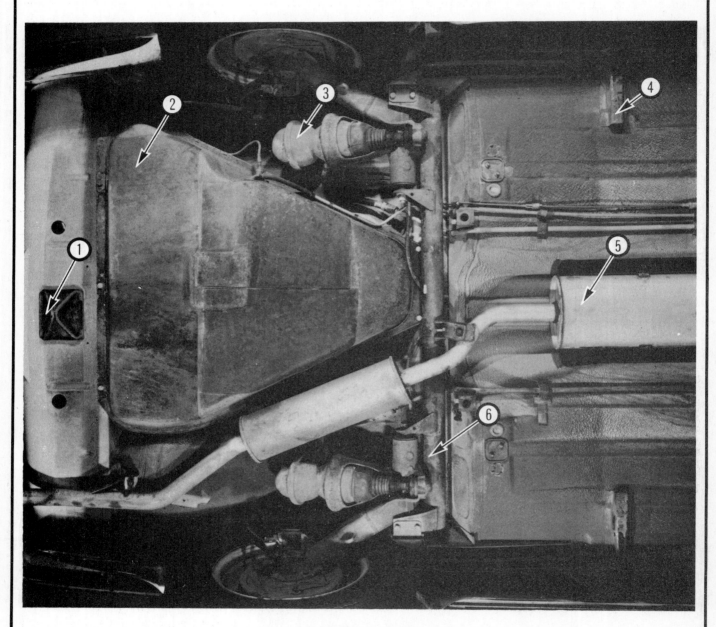

Underside of rear end

1	Rear towing hook	3	Displacer unit	5	Exhaust silencer
2	Fuel tank	4	Jacking point	6	Rear axle tube

Every 250 miles (400 km) or weekly

Check engine oil level
Check coolant level
Check windscreen washer fluid level
Check battery electrolyte level (not maintenance-free type batteries)
Check tyre pressures (including the spare)
Check the operation of all lights, indicators and horn
Check brake and clutch fluid levels

At first 1000 miles (1600 km) – new cars

Renew engine oil and filter
Check all brake, power steering and clutch hydraulic hoses and pipeline unions for leaks
Check tension of drivebelts
Tighten manifold nuts to the specified torque
Top up carburettor piston damper
Adjust engine idling speed and mixture (engine hot)
Check trim height (dealer operation if adjustment required)

Every 6000 miles (9600 km)

Renew engine oil and filter
Top up carburettor piston damper
Check and adjust idle speed and mixture
Check drivebelt tension
Check automatic transmission fluid level
Lubricate all control linkage, door hinges and locks
Clean and re-gap spark plugs
Check condition of contact breaker points and renew if necessary
Check and adjust dwell angle and ignition timing

Check power steering fluid level
Check tyres for wear and damage
Check brake pads and linings for wear
Check suspension bushes and steering balljoints for wear
Check condition of brake hydraulic hoses and pipes

Every 12 000 miles (19 000 km)

Renew engine oil filler/breather cap
Clean crankcase ventilation hoses
Check condition of driveshaft gaiters
Check condition of exhaust system
Check condition of cooling system hoses
Renew carburettor float chamber vent filter
Renew air cleaner element
Clean air intake filter of automatic starting unit (ASU)
Renew spark plugs
Check headlamp alignment
Check condition of steering rack gaiters
Check front wheel alignment
Check safety belts for wear or fraying
Grease the steering rack (early models)

Every 24 000 miles (38 000 km) or two yearly

Adjust toothed timing belt tension and check condition
Renew servo air filter
Renew antifreeze
Renew in-line fuel filter
Renew brake and clutch hydraulic fluid (by bleeding)

Every 48 000 miles (77 000 km) or four yearly

Renew toothed timing belt

Topping up engine oil

Topping up coolant

Checking a tyre pressure

Engine/transmission drain plug (manual)

Topping up power steering

Fault diagnosis

Introduction

The vehicle owner who does his or her own maintenance according to the recommended schedules should not have to use this section of the manual very often. Modern component reliability is such that, provided those items subject to wear or deterioration are inspected or renewed at the specified intervals, sudden failure is comparatively rare. Faults do not usually just happen as a result of sudden failure, but develop over a period of time. Major mechanical failures in particular are usually preceded by characteristic symptoms over hundreds or even thousands of miles. Those components which do occasionally fail without warning are often small and easily carried in the vehicle.

With any fault finding, the first step is to decide where to begin investigations. Sometimes this is obvious, but on other occasions a little detective work will be necessary. The owner who makes half a dozen haphazard adjustments or replacements may be successful in curing a fault (or its symptoms), but he will be none the wiser if the fault recurs and he may well have spent more time and money than was necessary. A calm and logical approach will be found to be more satisfactory in the long run. Always take into account any warning signs or abnormalities that may have been noticed in the period preceding the fault – power loss, high or low gauge readings, unusual noises or smells, etc – and remember that failure of components such as fuses or spark plugs may only be pointers to some underlying fault.

The pages which follow here are intended to help in cases of failure to start or breakdown on the road. There is also a Fault Diagnosis Section at the end of each Chapter which should be consulted if the preliminary checks prove unfruitful. Whatever the fault, certain basic principles apply. These are as follows:

Verify the fault. This is simply a matter of being sure that you know what the symptoms are before starting work. This is particularly important if you are investigating a fault for someone else who may not have described it very accurately.

Don't overlook the obvious. For example, if the vehicle won't start, is there petrol in the tank? (Don't take anyone else's word on this particular point, and don't trust the fuel gauge either!) If an electrical fault is indicated, look for loose or broken wires before digging out the test gear.

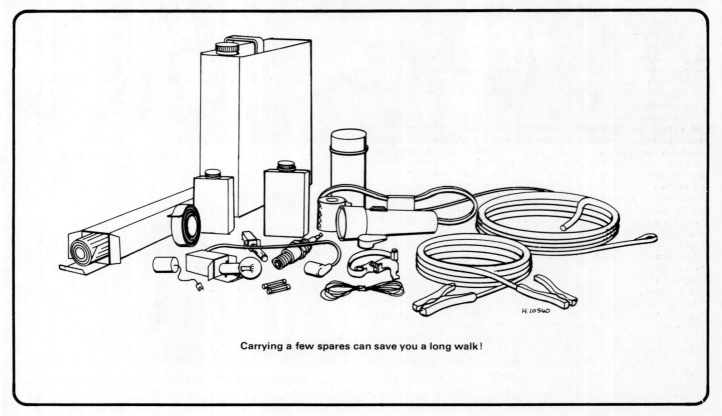

H.10540

Carrying a few spares can save you a long walk!

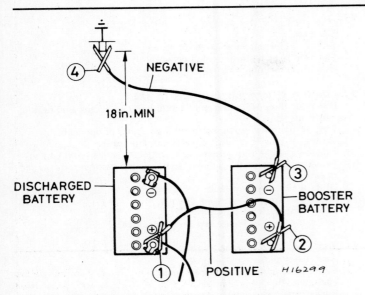

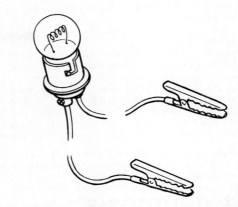

Jump start lead connections for negative earth vehicles – connect leads in order shown

A simple test lamp is useful for tracing electrical faults

Cure the disease, not the symptom. Substituting a flat battery with a fully charged one will get you off the hard shoulder, but if the underlying cause is not attended to, the new battery will go the same way. Similarly, changing oil-fouled spark plugs for a new set will get you moving again, but remember that the reason for the fouling (if it wasn't simply an incorrect grade of plug) will have to be established and corrected.

Don't take anything for granted. Particularly, don't forget that a 'new' component may itself be defective (especially if it's been rattling round in the boot for months), and don't leave components out of a fault diagnosis sequence just because they are new or recently fitted. When you do finally diagnose a difficult fault, you'll probably realise that all the evidence was there from the start.

Electrical faults

Electrical faults can be more puzzling than straightforward mechanical failures, but they are no less susceptible to logical analysis if the basic principles of operation are understood. Vehicle electrical wiring exists in extremely unfavourable conditions – heat, vibration and chemical attack – and the first things to look for are loose or corroded connections and broken or chafed wires, especially where the wires pass through holes in the bodywork or are subject to vibration.

All metal-bodied vehicles in current production have one pole of the battery 'earthed', ie connected to the vehicle bodywork, and in nearly all modern vehicles it is the negative (–) terminal. The various electrical components – motors, bulb holders etc – are also connected to earth, either by means of a lead or directly by their mountings. Electric current flows through the component and then back to the battery via the bodywork. If the component mounting is loose or corroded, or if a good path back to the battery is not available, the circuit will be incomplete and malfunction will result. The engine and/or gearbox are also earthed by means of flexible metal straps to the body or subframe; if these straps are loose or missing, starter motor, generator and ignition trouble may result.

Assuming the earth return to be satisfactory, electrical faults will be due either to component malfunction or to defects in the current supply. Individual components are dealt with in Chapter 10. If supply wires are broken or cracked internally this results in an open-circuit, and the easiest way to check for this is to bypass the suspect wire temporarily with a length of wire having a crocodile clip or suitable connector at each end. Alternatively, a 12V test lamp can be used to verify the presence of supply voltage at various points along the wire and the break can be thus isolated.

If a bare portion of a live wire touches the bodywork or other earthed metal part, the electricity will take the low-resistance path thus formed back to the battery: this is known as a short-circuit. Hopefully a short-circuit will blow a fuse, but otherwise it may cause

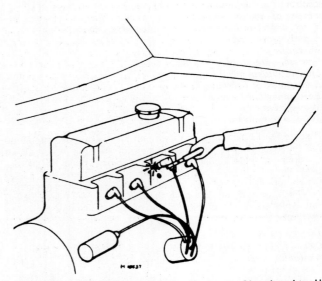

Crank engine and check for spark. Note use of insulated tool!

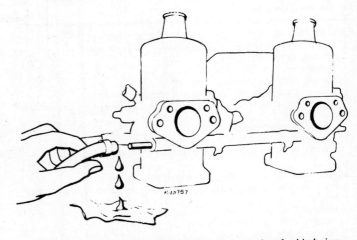

Remove fuel pipe from carburettor and check that fuel is being delivered

burning of the insulation (and possibly further short-circuits) or even a fire. This is why it is inadvisable to bypass persistently blowing fuses with silver foil or wire.

Spares and tool kit

Most vehicles are supplied only with sufficient tools for wheel changing; the *Maintenance and minor repair* tool kit detailed in *Tools and working facilities*, with the addition of a hammer, is probably sufficient for those repairs that most motorists would consider attempting at the roadside. In addition a few items which can be fitted without too much trouble in the event of a breakdown should be carried. Experience and available space will modify the list below, but the following may save having to call on professional assistance:

Spark plugs, clean and correctly gapped
HT lead and plug cap — long enough to reach the plug furthest from the distributor
Distributor rotor, condenser and contact breaker points
Drivebelt(s) — emergency type may suffice
Spare fuses
Set of principal light bulbs
Tin of radiator sealer and hose bandage
Exhaust bandage
Roll of insulating tape
Length of soft iron wire
Length of electrical flex
Torch or inspection lamp (can double as test lamp)
Battery jump leads
Tow-rope
Ignition waterproofing aerosol
Litre of engine oil
Sealed can of hydraulic fluid
Emergency windscreen
Worm drive clips
Tube of filler paste

If spare fuel is carried, a can designed for the purpose should be used to minimise risks of leakage and collision damage. A first aid kit and a warning triangle, whilst not at present compulsory in the UK, are obviously sensible items to carry in addition to the above.

When touring abroad it may be advisable to carry additional spares which, even if you cannot fit them yourself, could save having to wait while parts are obtained. The items below may be worth considering:

Throttle cable
Cylinder head gasket
Alternator brushes
Tyre valve core

One of the motoring organisations will be able to advise on availability of fuel etc in foreign countries.

Engine will not start

Engine fails to turn when starter operated
Flat battery (recharge, use jump leads, or push start)
Battery terminals loose or corroded
Battery earth to body defective
Engine earth strap loose or broken
Starter motor (or solenoid) wiring loose or broken
Automatic transmission selector in wrong position, or inhibitor switch faulty
Ignition/starter switch faulty
Major mechanical failure (seizure)
Starter or solenoid internal fault (see Chapter 10)

Starter motor turns engine slowly
Partially discharged battery (recharge, use jump leads, or push start)
Battery terminals loose or corroded
Battery earth to body defective
Engine earth strap loose
Starter motor (or solenoid) wiring loose
Starter motor internal fault (see Chapter 10)

Starter motor spins without turning engine
Flat battery
Starter motor pinion sticking on sleeve
Flywheel gear teeth damaged or worn
Starter motor mounting bolts loose

Engine turns normally but fails to start
Damp or dirty HT leads and distributor cap (crank engine and check for spark)
Dirty or incorrectly gapped distributor points (if applicable)
No fuel in tank (check for delivery at carburettor)
Excessive choke (hot engine) or insufficient choke (cold engine)
Fouled or incorrectly gapped spark plugs (remove, clean and regap)
Other ignition system fault (see Chapter 4)
Other fuel system fault (see Chapter 3)
Poor compression (see Chapter 1)
Major mechanical failure (eg camshaft drive)

Engine fires but will not run
Insufficient choke (cold engine)
Air leaks at carburettor or inlet manifold
Fuel starvation (see Chapter 3)
Ballast resistor defective, or other ignition fault (see Chapter 4)

Engine cuts out and will not restart

Engine cuts out suddenly — ignition fault
Loose or disconnected LT wires
Wet HT leads or distributor cap (after traversing water splash)
Coil or condenser failure (check for spark)
Other ignition fault (see Chapter 4)

Engine misfires before cutting out — fuel fault
Fuel tank empty
Fuel pump defective or filter blocked (check for delivery)
Fuel tank filler vent blocked (suction will be evident on releasing cap)
Carburettor needle valve sticking
Carburettor jets blocked (fuel contaminated)
Other fuel system fault (see Chapter 3)

Engine cuts out — other causes
Serious overheating
Major mechanical failure (eg camshaft drive)

Engine overheats

Ignition (no-charge) warning light illuminated
Slack or broken drivebelt — retension or renew (Chapter 2)

Ignition warning light not illuminated
Coolant loss due to internal or external leakage (see Chapter 1)
Thermostat defective
Low oil level
Brakes binding
Radiator clogged externally or internally
Electric cooling fan not operating correctly
Engine waterways clogged
Ignition timing incorrect or automatic advance malfunctioning
Mixture too weak

Note: *Do not add cold water to an overheated engine or damage may result*

Low engine oil pressure

Gauge reads low or warning light illuminated with engine running
Oil level low or incorrect grade

Defective gauge or sender unit
Wire to sender unit earthed
Engine overheating
Oil filter clogged or bypass valve defective
Oil pressure relief valve defective
Oil pick-up strainer clogged
Oil pump worn or mountings loose
Worn main or big-end bearings

Note: *Low oil pressure in a high-mileage engine at tickover is not necessarily a cause for concern. Sudden pressure loss at speed is far more significant. In any event, check the gauge or warning light sender before condemning the engine.*

Engine noises

Pre-ignition (pinking) on acceleration
Incorrect grade of fuel
Ignition timing incorrect
Distributor faulty or worn
Worn or maladjusted carburettor
Excessive carbon build-up in engine

Whistling or wheezing noises
Leaking vacuum hose
Leaking carburettor or manifold gasket
Blowing head gasket

Tapping or rattling
Incorrect valve clearances
Worn valve gear
Worn timing belt
Broken piston ring (ticking noise)

Knocking or thumping
Unintentional mechanical contact (eg fan blades)
Peripheral component fault (generator, coolant pump etc)
Worn big-end bearings (regular heavy knocking, perhaps less under load)
Worn main bearings (rumbling and knocking, perhaps worsening under load)
Piston slap (most noticeable when cold)

*It should be noted that with the type of fuel pump fitted to the Ambassador, switching on the ignition actuates the fuel pump relay, but fuel will not actually be ejected from the outlet pipe of the pump until the ignition key is turned to the start position.

Chapter 1 Engine

Contents

Specifications

Engine type .... Four-cylinder in-line overhead camshaft. Transversely mounted over transmission

Designation and capacity
17H .... 103.73 in^3 (1700 cc)
20H .... 121.68 in^3 (1994 cc)

Bore .... 3.325 in (84.45 mm)

Stroke
17H .... 2.984 in (75.8 mm)
20H .... 3.504 in (89.0 mm)

Firing order .... 1 – 3 – 4 – 2 (No 1 at crankshaft pulley end)

Compression ratio 9.0:1

Cylinder compression .... 170 to 195 lbf/in^2 (11.9 to 13.7 bar)

Maximum power (DIN)
17H .... 83 bhp at 5200 rev/min
20H .... 92 bhp at 4900 rev/min

Maximum torque
17H .... 97 lbf ft at 3500 rev/min
20H .... 114 lbf ft at 2750 rev/min

Crankshaft
Main journal diameter .... 2.1262 to 2.1270 in (54.005 to 54.026 mm)
Crankpin diameter .... 1.8754 to 1.8759 in (47.635 to 47.647 mm)
Crankshaft endfloat .... 0.001 to 0.005 in (0.025 to 0.13 mm)
Number of main bearings 5
Running clearance .... 0.001 to 0.003 in (0.025 to 0.077 mm)
Big-end bearing running clearance .... 0.0015 to 0.0032 in (0.038 to 0.081 mm)

Pistons
Type .... Solid skirt with recess in crown
Number of piston rings .... 2 compression, 1 oil control
Ring to groove clearance (compression rings) .... 0.0015 to 0.0027 in (0.04 to 0.07 mm)
End gap (fitted in bore):
    Compression rings .... 0.012 to 0.020 in (0.3 to 0.5 mm)
    Oil control ring (no gap in expander) .... 0.015 to 0.055 in (0.38 to 1.4 mm)
Gudgeon pin type .... Interference fit in connecting rod

Camshaft
Endfloat .... 0.003 to 0.007 in (0.07 to 0.18 mm)
Journal diameters .... 1.888 to 1.889 in (47.96 to 47.97 mm)
Bearings .... 3, direct in cylinder head and cam cover
Running clearance .... 0.0017 to 0.0037 in (0.043 to 0.094 mm)

Timing belt
Type .... Flexible toothed
Number of teeth .... 104 x 0.375 in pitch (9.52 mm)

Tappets and shims
Type .... Inverted bucket tappet with shim located in valve spring cap recess

Valves
Face angle .... 45° 30'
Seat angle .... 45°
Head diameter:
    Inlet .... 1.575 in (40.0 mm)
    Exhaust .... 1.339 in (34.0 mm)
Stem diameter:
    Inlet .... 0.2917 to 0.2921 in (7.41 to 7.42 mm)
    Exhaust .... 0.2909 to 0.2917 in (7.39 to 7.41 mm)
Stem to guide clearance:
    Inlet .... 0.001 to 0.002 in (0.025 to 0.053 mm)
    Exhaust .... 0.0015 to 0.0028 in (0.04 to 0.073 mm)
Cam lift .... 0.375 in (9.525 mm)

Valve guides
Length .... 1.532 in (38.90 mm)
Outside diameter .... 0.474 to 0.475 in (12.04 to 12.06 mm)
Inside diameter .... 0.2933 to 0.2937 in (7.45 to 7.46 mm)
Interference fit in cylinder head .... 0.0015 to 0.003 in (0.04 to 0.09 mm)
Projection of guide above cylinder head .... 0.394 in (10.0 mm)

Valve springs
Free length .... 1.646 in (41.81 mm)
Fitted length .... 1.375 in (34.92 mm)

Valve clearance
Standard .... 0.017 to 0.019 in (0.43 to 0.48 mm)
Service adjustment only if less than .... 0.008 in (0.20 mm)

Valve timing

	17H	20H
Inlet valve:		
Opens	15° BTDC	19° BTDC
Closes	45° ABDC	41° ABDC
Exhaust valve:		
Opens	50° BBDC	61° BBDC
Closes	10° ATDC	15° ATDC

Lubrication

System	Wet sump with rotor type pump on front of crankshaft
Oil pressure:	
Idling	40 to 60 lbf/in² (2.8 to 4.1 bar)
At road speed	60 to 90 lbf/in² (4.1 to 6.2 bar)
Relief valve spring free length	1.525 in (38.7 mm)
Lubricant type/specification*	Multigrade engine oil, viscosity SAE 10W/40, 15W/40 or 15W/50, to API SE/SF (Duckhams QXR, Hypergrade or 10W/40 Motor Oil)
Oil capacity with manual transmission (joint oil supply):	
Oil change with new filter	10.25 pts (5.8 litres)
Oil capacity with automatic transmission (independent oil supply):	
Oil change with new filter	6.50 pts (3.7 litres)

*Note: Austin Rover specify a 10W/40 oil to meet warranty requirements for models produced after August 1983. Duckhams QXR and 10W/40 Motor Oil are available to meet these requirements.

Torque wrench settings

	lbf ft	Nm
Cylinder head bolts:		
Stage 1	35	41
Stage 2	65	88
Stage 3	Release each bolt (¼ turn) in sequence and retighten to 65 lbf ft (88 Nm)	
Camshaft rear end cover bolts	8	11
Camshaft cover bolts	13	18
Distributor mounting flange screws	18	25
Camshaft sprocket bolt	35	41
Manifold bolts	18	25
Carburettor mounting nuts	19	26
Air cleaner bolts	19	26
Crankshaft pulley bolt	65	88
Main bearing cap bolts	75	102
Big-end cap nuts	32	44
Flywheel mounting screws	42	57
Clutch cover screws	17	23
Oil pump mounting bolts	8	11
Coolant pump bolts	8	11
Thermostat housing bolts	8	11
Cylinder block coolant drain plug	27	37
Engine mounting (long through-bolt) locknut	30	41
For setscrews not specified entering tapped holes in aluminium:		
Size M6	8	11
Size M8	18	25
For setscrews not specified entering tapped holes in cast-iron:		
Size M6	8	11
Size M8	22	30
Size M10	37	50

For manual gearbox and automatic transmission torque wrench settings, refer to Specifications Section in Chapter 6.

1 General description

The O-Series engine used to power the models described in this manual is of four cylinder overhead camshaft (ohc) type, mounted transversely at the front of the car.

The cylinder block is of cast-iron while the cylinder head is of light alloy.

The camshaft is driven by a toothed belt and an unusual feature is that the cam cover also acts as the camshaft upper bearing sections.

The crankshaft runs in five main bearings and the oil pump is mounted on its front end.

The connecting rods are of forged steel having gudgeon pins which are an interference fit. No retaining circlips in the piston bosses are therefore required.

The valve clearance is set by shims and any adjustment should be required infrequently as generous tolerances are specified.

The camshaft is used to drive the distributor by means of a gear.

The engine has a mechanical breaker type ignition distributor and inlet and exhaust manifolds on the same side of the cylinder head. Apart from an instantly removable cooling system thermostat, the engine maintains a conventional design characteristic.

On cars with manual transmission, the engine oil is also used jointly with the transmission and is contained in the gearcasing.

2 Engine oil and filter

1   It is important that the engine oil level is checked at weekly intervals in the following way.
2   Make sure that the dipstick is fully inserted into its guide tube with the word OIL on its moulded handle readable when standing in front of the car.
3   With the engine either cold or having been switched off for at least five minutes, withdraw the dipstick, wipe it clean and reinsert it (OIL again readable from front of car).
4   Withdraw it for the second time and read off the oil level from its upward facing surface (marked MIN and MAX).
5   Top up with specified oil to the MAX notch on the dipstick. The amount of oil required to raise the oil level from the MIN to MAX marks is 1.5 pints (0.9 litre).
6   When draining the engine oil, it is preferable to do this when it is hot.
7   Unscrew the drain plug noting the different location depending

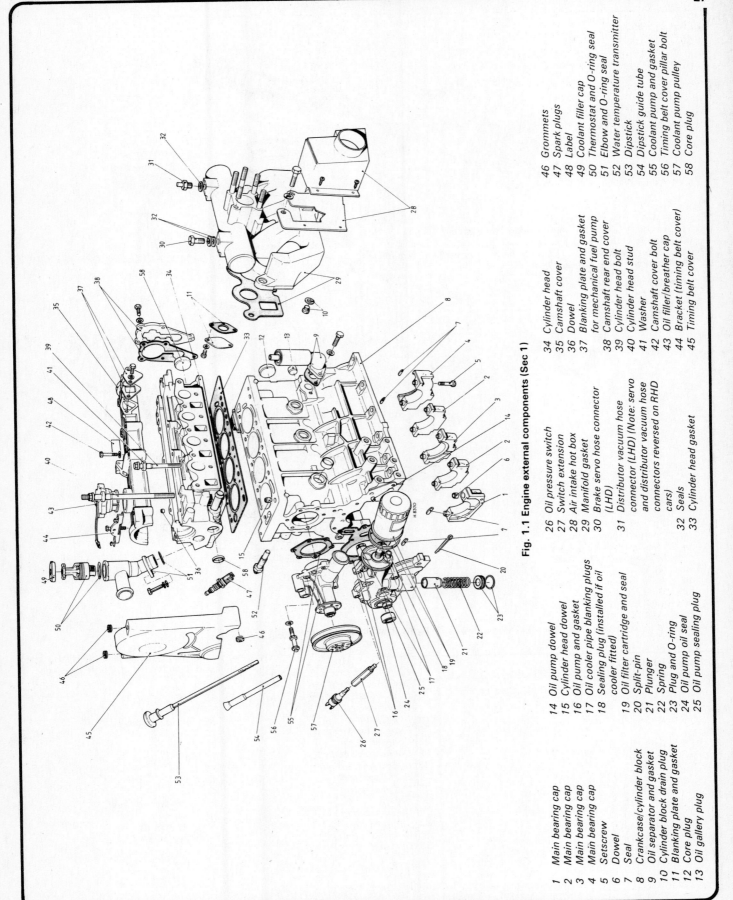

Fig. 1.1 Engine external components (Sec 1)

1 Main bearing cap
2 Main bearing cap
3 Main bearing cap
4 Main bearing cap
5 Setscrew
6 Dowel
7 Seal
8 Crankcase/cylinder block
9 Oil separator and gasket
10 Cylinder block drain plug
11 Blanking plate and gasket
12 Core plug
13 Oil gallery plug

14 Oil pump dowel
15 Cylinder head dowel
16 Oil pump and gasket
17 Oil cooler pipe blanking plugs
18 Sealing plug (installed if oil cooler fitted)
19 Oil filter cartridge and seal
20 Split-pin
21 Plunger
22 Spring
23 Plug and O-ring
24 Oil pump oil seal
25 Oil pump sealing plug

26 Oil pressure switch
27 Switch extension
28 Air intake hot box
29 Manifold gasket
30 Brake servo hose connector (LHD)
31 Distributor vacuum hose connector (LHD) (Note: servo and distributor vacuum hose connectors reversed on RHD cars)
32 Seals
33 Cylinder head gasket

34 Cylinder head
35 Camshaft cover
36 Dowel
37 Blanking plate and gasket for mechanical fuel pump
38 Camshaft rear end cover
39 Cylinder head bolt
40 Cylinder head stud
41 Washer
42 Camshaft cover bolt
43 Oil filler/breather cap
44 Bracket (timing belt cover)
45 Timing belt cover

46 Grommets
47 Spark plugs
48 Label
49 Coolant filler cap
50 Thermostat and O-ring seal
51 Elbow and O-ring seal
52 Water temperature transmitter
53 Dipstick
54 Dipstick guide tube
55 Coolant pump and gasket
56 Timing belt cover pillar bolt
57 Coolant pump pulley
58 Core plug

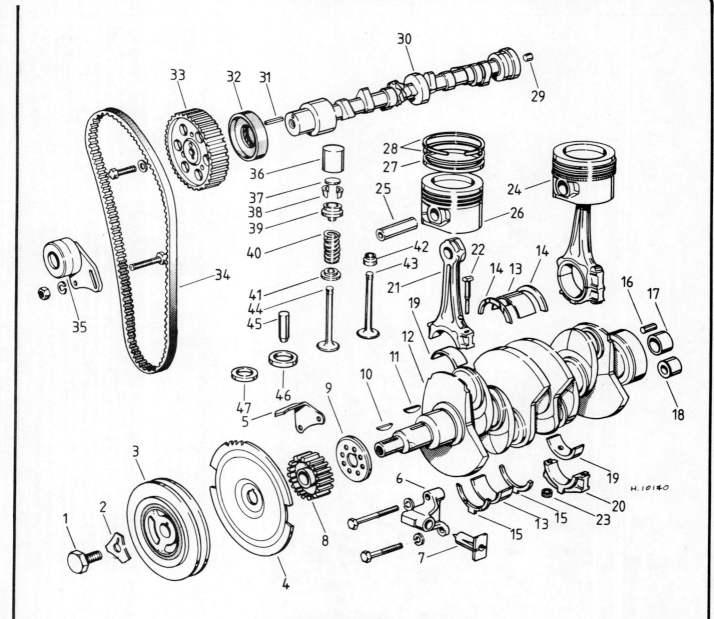

Fig. 1.2 Engine internal components (Sec 1)

1 Crankshaft pulley bolt	17 Spigot bush (manual gearbox 1st motion shaft)	31 Roll pin (dowel)
2 Lockplate		32 Oil seal
3 Pulley vibration damper	18 Spigot bush (automatic transmission)	33 Timing sprocket
4 Timing disc		34 Timing belt
5 Pointer	19 Shell bearing	35 Tensioner pulley
6 Bracket (LED ignition timing)	20 Big-end bearing cap	36 Tappet
7 Plug	21 Connecting rod	37 Shim
8 Timing sprocket	22 Connecting rod bolt	38 Split cotters
9 Flange plate	23 Connecting rod nut	39 Cup
10 Woodruff key	24 Piston assembly	40 Spring
11 Woodruff key	25 Gudgeon pin	41 Seat
12 Crankshaft	26 Piston	42 Valve stem oil seal
13 Shell bearings	27 Oil control ring	43 Inlet valve
14 Thrust washer	28 Compression ring	44 Exhaust valve
15 Thrust washer	29 Camshaft plug	45 Valve guide
16 Dowel	30 Camshaft	46 Valve seat insert
		47 Valve seat insert

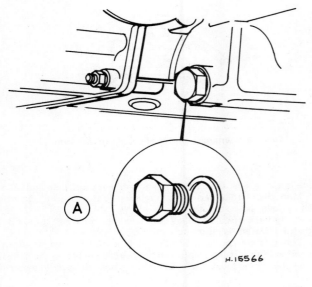

2.10A Oil filter being fitted

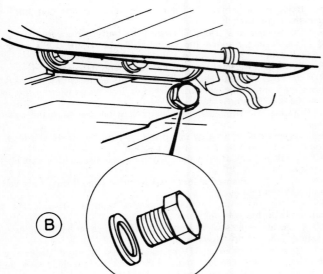

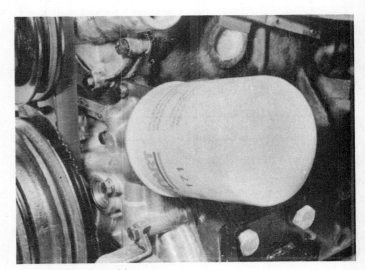

2.10B Oil filter in position

H.15566

Fig. 1.3 Engine oil drain plugs (Sec 2)

A   *Manual transmission*   B   *Automatic transmission*

upon whether the car has manual or automatic transmission, and catch the oil in a suitable container.

8   Wipe the drain plug magnet clean and once the oil has finished draining, screw the plug back tightly.

9   Unscrew the cartridge type oil filter and discard it. A filter removal tool may be needed.

10   Smear some grease on the rubber seal of the new filter and screw it home hand tight only (photos).

11   Fill the engine with good quality fresh oil of the specified grade.

12   Start the engine. A few seconds will elapse before the oil pressure warning lamp goes out. This is normal and is due to the new filter having to fill with oil.

13   Check the new oil filter gasket for leaks, switch off and after a few minutes, check the oil level and top up if necessary.

3   Crankcase ventilation system

1   A very elementary type of crankcase ventilation system is used incorporating a filter in the oil filler cap through which air is drawn to replace the blow-by gas and oil fumes which are extracted through an oil separator and hose to the carburettor (photo).

3.1 Crankcase ventilation separator and hose

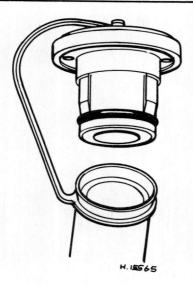

Fig. 1.4 Engine oil filler plug and crankcase breather (inlet) (Sec 3)

2   Clean the hose and separator also the hose restrictor (if fitted) and renew the oil filler cap complete at the specified intervals (see Routine Maintenance at the beginning of this Manual).

4   Major operations possible without removing engine from car

The following work can be carried out while the engine is in position in the car:

(a) Removal and refitting of the camshaft
(b) Renewal of the crankshaft front oil seal
(c) Removal, refitting and tensioning of the timing belt
(d) Removal and refitting of the cylinder head
(e) Removal and refitting of the oil pump
(f) Renewal of the engine flexible mountings
(g) Valve clearance adjustment

5   Camshaft – removal and refitting

1   Remove the timing belt cover, by pulling it off its rubber cleats.
2   Disconnect the spark plug leads, pull off the distributor cap and

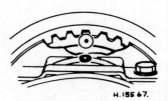

Fig. 1.5 Camshaft sprocket timing marks (Sec 5)

place it to one side. Disconnect the advance vacuum pipe and LT lead.
3   Mark the distributor setting in relation to the camshaft cover and then remove the distributor.
4   Using the crankshaft pulley bolt, turn the crankshaft to 90° BTDC with No 1 piston rising on its compression stroke.
5   Continue turning until the V-notch in the crankshaft pulley is aligned with the timing pointer, and the single dimple on the camshaft sprocket is aligned with the mark on the camshaft cover (viewed from rear).
6   Release the timing belt tensioner and slip the belt from the camshaft sprocket.
7   Do not turn the crankshaft while the belt is removed.
8   Unbolt and remove the camshaft sprocket. Hold the sprocket against rotation by passing a rod through one of its holes.
9   Remove the camshaft rear end cover and its gasket.
10 Unscrew the nine camshaft cover bolts a turn at a time, working from the centre ones towards each end until the tension of the valve springs is relieved. Remove the cover.
11 Lift the camshaft complete with front oil seal from the cylinder head.
12 Commence refitting by oiling the camshaft journals and locating the camshaft in its cylinder head bearings.
13 Check and adjust the valve clearances as described in Section 11, paragraphs 10 to 26.
14 Locate the selected shims in their spring cup on top of the valves and refit the tappets.
15 Oil the camshaft journals and the tappets, then locate the camshaft on the cylinder head (photo).
16 Clean away old sealant from the camshaft cover and cylinder head mating flanges and wipe clean.
17 Apply a thin bead of RTV silicone sealant all round the centre of the flange on the cylinder head and then bolt the camshaft cover down evenly and progressively. Remember that the timing belt cover bracket is located under the front two bolts (photo).
18 Fit the camshaft end cover with a new gasket (photo).
19 Grease the lip and outer edge of the new camshaft oil seal and drive it into position, using a hammer and a piece of hardwood (photo).

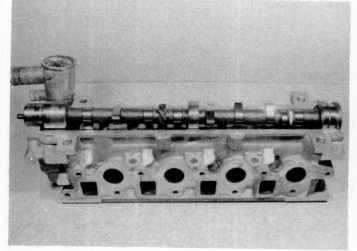

5.15 Camshaft on cylinder head

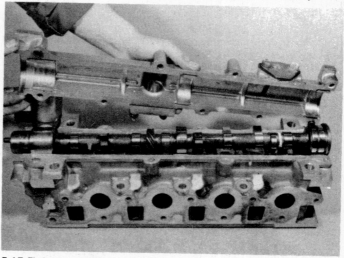

5.17 Fitting camshaft cover

5.18 Camshaft end cover and gasket

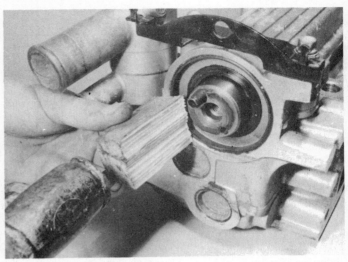

5.19 Fitting camshaft oil seal

20 Refit the camshaft sprocket, tightening the bolt to the specified torque. The sprocket is dowelled and will only fit one way (photo).
21 Engage the timing belt with the camshaft sprocket and tension the belt (Section 7).
22 Turn the crankshaft until, with No 1 piston rising on compression, the 8° BTDC timing mark on the pulley is aligned with the pointer.
23 Fit the distributor as described in Chapter 4.
24 Connect the distributor cap, spark plug leads advance vacuum pipe and LT lead.
25 Refit the timing belt cover.
26 Adjust the ignition timing as described in Chapter 4 using a stroboscope.

6 Crankshaft front oil seal – renewal

1 Remove the camshaft timing belt cover.
2 Remove the power steering pump and alternator drivebelts as described in Section 7, paragraphs 2 and 3.
3 Unbolt and remove the power steering pulley from the crankshaft pulley.
4 Bend back the tab washer and slacken the crankshaft pulley retaining bolt but do not remove it yet. The bolt may release if a short ring spanner or socket with short bar is used and the bar struck a sharp blow with a hammer. It is useless attempting to unscrew the bolt in the normal way unless the flywheel ring has been jammed to prevent the crankshaft rotating in the reverse direction of travel. To jam the ring gear, remove the starter motor (Chapter 10) and place a cold chisel or piece of flat steel between a tooth on the ring gear and the edge of the starter motor aperture.
5 If access to the crankshaft pulley is better from above, release the coolant pipe support bracket and pull the pipe to one side out of the way.
6 Before the crankshaft pulley bolt is removed, set the crankshaft to 90° BTDC and the camshaft sprocket as described in paragraph 5 of Section 5.
7 Unscrew the crankshaft pulley bolt, remove it with lockplate and then pull off the pulley. If it is tight, lever it off gently by inserting two levers, each at opposite points of the pulley, so that pressure can be applied as near the centre of the pulley as possible to prevent undue strain on the rubber damper bonding.
8 Extract the plug from the LED optical sensor bracket (see Chapter 4).
9 Remove the timing disc.
10 Release the timing belt tensioner and withdraw the timing belt.
11 From the front end of the crankshaft, pull off the timing belt sprocket leaving the belt guide plate at its rear in position as the plate

5.20 Tightening camshaft sprocket bolt

keyway does not engage with the same Woodruff key as the sprocket and is unlikely to be in correct alignment for removal. Once the sprocket has been removed, withdraw the plate.
12 Extract the Woodruff key from the crankshaft.
13 Carefully prise out the oil seal from the oil pump housing. Use a hooked tool or two-legged outward facing claw extractor for this, taking care not to scratch the crankshaft surface.
14 Oil the new seal and drive it into position using a piece of tubing.
15 Refit all the components by reversing the removal operations but note the following points:

(a) *Make sure that the crankshaft and camshaft have not moved from their originally aligned positions*
(b) *Fit the belt guide plate to the crankshaft making sure that its convex side is visible, then fit the sprocket so that its semi-circular depression is against the plate*
(c) *Fit the timing disc*
(d) *Fit the pulley*
(e) *Use a new pulley bolt lockplate*
(f) *Tighten all bolts to the specified torque and tension the timing belt as described in Section 7 paragraphs 11 and 12 and the drivebelts as described in Section 7 paragraph 13*

7 Timing belt – removal, refitting and tensioning

1   The timing belt should be inspected at regular intervals for cracks, fraying and the rounding off of the teeth. The belt must be renewed if these conditions are evident.

2   As the car is equipped with power steering, slacken the rigid pipe union on the steering pump, slacken the pump mounting bolts, push the pump in towards the engine and slip the drivebelt from the pump and crankshaft pulleys.

3   Slacken the alternator mounting bolts and adjuster link bolt, push the alternator in towards the engine and slip the drivebelt from the alternator and crankshaft pulleys.

4   Turn the crankshaft to 90° BTDC (single V-notch in crankshaft pulley timing disc opposite pointer). *This operation is very important* to prevent the valves impinging on the piston crowns should the crankshaft or camshaft be turned while the timing belt is off (photo).

5   Remove the timing belt cover by pulling its spigots out of their rubber grommets and then release the timing belt tensioner pulley.

6   Ease the timing belt off the camshaft gear, pull the plug from LED optical sensor bracket (see Chapter 4)

7   Turn the crankshaft pulley until the timing belt can be removed from the crankshaft sprocket through the cut-out provided in the timing disc.

8   Before fitting the new belt, make sure that the belt, gear teeth and tensioner pulley are all free from oil and dirt.

9   Fit the new belt to the crankshaft sprocket again passing it through the cut-out in the timing disc.

10   Before engaging the timing belt with the camshaft sprocket, turn the crankshaft pulley to align the V-notch in the timing disc with the timing pointer (90° BTDC). Then check that the camshaft sprocket is still correctly aligned with the dimple on its rear face opposite the pointer on the camshaft cover.

11   Engage the timing belt with the camshaft sprocket and then tension it. To do this, attach a spring balance to the belt in line with the coolant inlet pipe of the coolant pump.

12   The pull now required to bring the smooth side of the timing belt in alignment with the moulded line on the inlet pipe should be 11 lbf

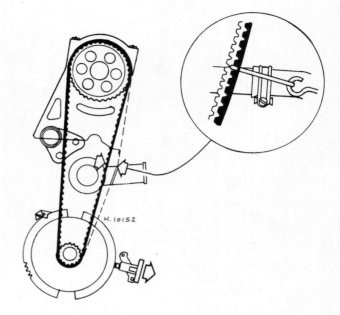

Fig. 1.6 Timing belt tensioning diagram (Sec 7)

(5 kgf) for a used belt or 13 lbf (6 kgf) for a new belt. Adjust the tensioner pulley as necessary to achieve this (photos).

13   Fit the timing belt cover by pushing its spigots into the rubber cleats (photos).

14   Refit the drivebelts and tension them as described in Chapter 2, Section 10.

7.4 Crankshaft timing disc at 90° BTDC

7.12A Checking belt tension

7.12B Timing belt tensioner

7.13A Timing belt cover lower cleat

7.13B Timing belt upper fixing spigots

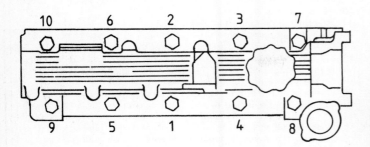

Fig. 1.7 Cylinder head bolt tightening sequence (Sec 8)

8.9 Cylinder head gasket

8 Cylinder head – removal and refitting

1   *If the engine is in the car,* carry out the following preliminary operations:

(a)   *Disconnect the battery and the leads from the temperature transmitter, spark plugs and coil negative terminal*
(b)   *Drain the coolant (see Chapter 2)*
(c)   *Remove the air cleaner (see Chapter 3)*
(d)   *On cars with left-hand steering, remove the throttle and choke control cable straps*
(e)   *Disconnect the distributor vacuum pipe*
(f)   *Remove the timing belt cover*
(g)   *Disconnect the brake servo vacuum pipe from the intake manifold and the breather hose from the carburettor*
(h)   *Disconnect the fuel pipe from the carburettor and then unbolt the carburettor and place it to one side. On cars equipped with automatic transmission, the downshift cable will have to be disconnected before the carburettor can be moved. Remove the heat shield and flange insulator*
(j)   *Disconnect the exhaust pipes from the manifold*
(k)   *Disconnect the heater inlet hose and the radiator top hose from the cylinder head*
(l)   *Unbolt the power steering pump bracket*

2   Set the crankshaft and camshaft as described in Section 5.
3   Release the timing belt tensioner and ease the timing belt from the camshaft sprocket.
4   With all attachments now removed from the cylinder head and camshaft cover, unscrew the cylinder head bolts, a turn at a time in diagonally opposite sequence starting with the two centre bolts and working towards each end of the cylinder head. A hexagon (Allen key) type socket is required.
5   With the help of an assistant, lift the cylinder head complete with cam cover and camshaft from the block. If it is stuck, lever the head under the two lugs on the spark plug side. **Do not** insert a lever elsewhere or the mating surfaces will be damaged. An obstinate cylinder head can normally be freed if the manifolds are gripped and a rocking motion applied to them. Do not tap the cylinder head sideways as it is located on two hollow dowels.
6   Peel away the cylinder head gasket.
7   Before commencing refitting, remove all the old gasket and carbon from the head and block (see Section 18).
8   Mop out any oil and dirt from the bolt holes in the block. This is important to prevent hydraulic pressure cracking the block when the bolts are screwed in.
9   Position a new gasket on the cylinder block. Make sure that the cylinder block surface is perfectly clean and fit the gasket dry (photo).
10   Lower the cylinder head into position over its dowels. Smear the bolt threads with engine oil and screw them in finger tight. Tighten the bolts in the stages given in 'Torque Wrench Settings' in the Specifications working in the sequence shown in Fig. 1.7 (photos).
11   *If the cylinder head was not dismantled,* make sure that the crankshaft and camshaft settings are still as described in paragraph 10 of Section 7 and refit the timing belt.
12   Tension the timing belt as described in Section 7.
13   Reverse the operations described in paragraph 1.

8.10A Lowering cylinder head onto block

8.10B Tightening a cylinder head bolt

9   Oil pump – removal and refitting

1   Remove the camshaft timing belt cover, the engine drivebelts, the crankshaft pulley and the crankshaft timing sprocket, all as described in Section 6.
2   Remove the Woodruff key from the crankshaft.
3   Disconnect the leads from the oil pressure switch and remove the oil filter cartridge.
4   Where fitted, disconnect the oil cooler hoses from the oil pump body, noting that the return hose is the one at the bottom. Tie the hoses up to prevent loss of oil.
5   Slacken the nine screws which hold the oil pump housing to the cylinder block.
6   Unbolt the LED optical sensor bracket noting that the longest screw is at the top.
7   Remove the oil pump housing screws noting that the long screw is adjacent to the relief valve plug.

8   Withdraw the oil pump and be prepared to catch some oil which will be released.
9   Peel off the oil pump flange gasket.
10 With the oil pump removed, clean away all external dirt.
11 Fit a new crankshaft oil seal to the pump.
12 If a new oil pump is being fitted check that the oilway is clear if no engine oil cooler is fitted or plugged if an engine oil cooler is fitted.
13 Fit the Woodruff key to the crankshaft (photo).
14 Fit the oil pump to the engine by reversing the removal operations described at the beginning of this Section. Always use a new flange gasket. Tape the crankshaft to prevent damage to the oil seal (photo).
15 Once the oil pump is bolted into place, prime it with engine oil injected into the lower oil cooler hose hole or plug hole (photos).
16 Refit the timing components, crankshaft pulley and drivebelts, as described in Section 7.
17 Start the engine and check that the oil pressure warning lamp goes out. Inspect disturbed components for leaks. Top-up the engine oil.

9.13 Crankshaft Woodruff key

9.14 Crankshaft shoulder taped

9.15A Tightening oil pump bolt

9.15B Oil pump fitted

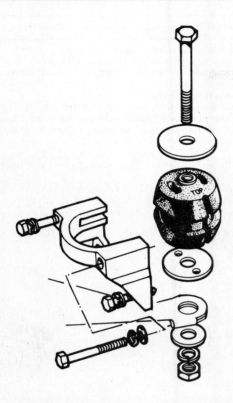

Fig. 1.8 Typical engine mounting components (Sec 10)

10 Engine mountings – renewal

1   A trolley jack will be required to support the weight of the engine/gearbox during the following operations. Position the jack under the gearbox and, using a block of wood as an insulator, raise the jack just enough to take the weight of the engine/gearbox.

2   Remove the long bolt which passes through the flexible mounting cushion.

3   Slightly raise or lower the engine as necessary to be able to unscrew the mounting bracket bolts.

4   Withdraw the bracket and extract the flexible cushion (Fig. 1.8).

5   If more than one of the engine mountings is being removed, take

care not to mix up the components, which differ between front and rear.

6   To refit a mounting, locate the flexible cushion in its support bracket so that the flat is towards the car body. The cushions are marked TOP and are fitted as colour matched sets of two according to whether they are rear (blue) or front (orange).

7   Fit the lower plate over the spikes which are moulded onto the cushion. Once the plate is a snug fit, cut off the spikes flush with the surface of the plate.

8   Refit the components by reversing the removal operations, noting that on the right-hand front mounting, a large steel washer is located under and above the flexible cushion. Tighten all bolts to the specified torque.

11 Valve clearances – adjustment

1   Remove the timing belt cover.

2   Unscrew the holding screws from the distributor cap, pull off the cap and place it to one side.

3   Disconnect the vacuum pipe from the distributor.

4   Mark the relative position of the distributor mounting flange to the cam cover for ease of refitting. Unscrew the fixing bolts and withdraw the distributor.

5   Turn the crankshaft to the 90° BTDC position on the compression stroke. This is achieved when the single dimple on the rear face of the camshaft sprocket is opposite the pointer on the front of the upper surface of the cam cover. At the same time the V-notch in the crankshaft pulley timing disc is opposite the timing pointer. The crankshaft can be turned using a socket on the crankshaft pulley bolt or, on cars with manual transmission only, by jacking up a front roadwheel, engaging top gear and turning the wheel in the normal forward direction of rotation. Rotation of the crankshaft will be made easier if the spark plugs are first removed (photo).

6   Release the timing belt tensioner pulley and prise the timing belt from the camshaft sprocket.

7   Unbolt and remove the camshaft sprocket.

8   Unbolt and remove the camshaft rear end cover with its gasket.

9   Unscrew each of the nine cam cover bolts evenly and in side-to-side sequence until the pressure of the valve springs on the cam cover can be felt to be relieved. Lift the cam cover away.

10   Three clamps will have to be fitted to keep the camshaft depressed against the pressure of the valve springs. These clamps (Tool no 18G1301) are expensive and their use is not essential if the following procedure is followed precisely. Make up some short lengths of angle iron or similar and bolt them to the cylinder head using the cylinder head studs or bolts. Make sure that the underside of the clamp is insulated with a strip of wood or other material to protect the surface of the camshaft bearings and lobes (photo).

11   Fit the clamps and tighten to relieve the pressure of the valve springs until the camshaft is fully seated in its bearings with Nos 1 and

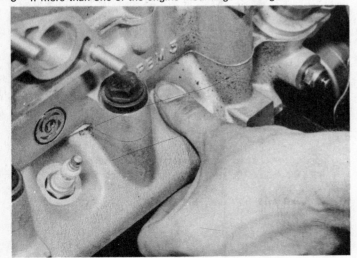

11.5 Feeling compression on No. 1 cylinder

11.10 Camshaft clamp

4 cam lobes having their peaks 45° either side of vertical (No 1 at timing belt end).

12 Insert the appropriate feeler blade (see Specification Section) between the cam lobe and the tappet (cam follower). Record the clearance.

13 Now check the clearance at No 4 cam lobe.

14 Slacken the clamps and turn the camshaft to bring the next pair (2 and 5) of cam lobes into position 45° either side of vertical.

15 Tighten the clamps and check the clearances and record them.

16 Repeat the operations on the remaining valves in the sequence 6 and 7 then 3 and 8.

17 Compare the recorded clearances with those specified at the beginning of this Chapter. The clearances need only be adjusted if they are less than 0.008 in (0.20 mm).

18 If any or all valve clearances are outside the specified tolerance, remove the clamps and lift off the camshaft.

19 Withdraw the tappet (cam follower) and extract the shim from its recess in the valve spring cap.

20 *It is very important during this work not to mix up the tappets or shims* but to keep them in their originally fitted sequence.

21 Use a tray, sub-divided into eight components, for keeping the shims and tappets in order. They should be numbered 1 to 8 starting from the timing sprocket end of the camshaft.

22 Measure the thickness of the shim with a micrometer and by simple calculation, note the thickness of the new shim which must be obtained to increase or decrease the clearance and bring it within tolerance.

23 Shims are available in the following thicknesses:

in	mm
0.091	2.32
0.093	2.37
0.095	2.42
0.097	2.47
0.099	2.52
0.101	2.56
0.103	2.62
0.105	2.67
0.107	2.72
0.109	2.77
0.111	2.83
0.113	2.87
0.115	2.93

It is sometimes possible to interchange two shims from tappets which have incorrect clearances so that they come within tolerance without the need to buy new shims.

24 Fit the tappet shims into their recesses in the valve spring caps. Insert the tappets into their guides.

25 Oil the camshaft journals and tappets and locate the camshaft on the cylinder head.

26 Refit the cam cover. To do this, peel away the old gasket joint, wipe the mating faces clean and apply a new bead $\frac{1}{16}$th in (1.5 mm) diameter of RTV silicone sealant. Fit the cover immediately and tighten the cover bolts evenly and progressively, in side-to-side sequence, to the specified torque.

27 Fit the bracket for the timing belt cover and the camshaft rear end cover and gasket.

28 Fit a new camshaft oil seal fully into its housing and grease its lips.

29 Fit the camshaft sprocket and tighten its retaining bolt to the specified torque.

30 Check that the crankshaft is set to the 90° BTDC position as described in paragraph 5.

31 Set the camshaft so that the dimple on the rear face of the sprocket is also aligned as described in paragraph 5.

32 Fit the timing belt and tension it as described in Section 7.

33 Refit the distributor as described in Chapter 4. Reconnect the vacuum pipe.

34 Refit the timing belt cover.

35 Adjust the ignition timing as described in Chapter 4.

12 Engine – method of removal

1 The engine should be removed complete with transmission by hoisting it upward out of the engine compartment.

13 Engine/manual transmission – removal and separation

1 Disconnect and remove the battery.

2 Drain the cooling system as described in Chapter 2.

3 Disconnect the bottom coolant hose from the coolant pump and from its connecting tube.

4 Remove the timing belt cover.

5 Disconnect the heater hose from the cylinder head (photo).

6 Disconnect the radiator top hose from the thermostat housing.

7 Remove the radiator (Chapter 2) and the expansion tank (photo).

13.5 Heater hose at cylinder head

13.7 Expansion tank mounting frame screw

8 Disconnect the horn leads and remove the battery tray and carrier.

9 Disconnect all leads from the engine including those to the following:

Starter solenoid (photo)
Ignition coil (photo)
Coolant temperature transmitter (photo)
Oil pressure switch (photo)
Alternator. Release the wiring harness clips (photo)

10 Remove the air cleaner and air temperature control assembly.

11 Unscrew and remove the brake servo connector from the intake manifold (photo).

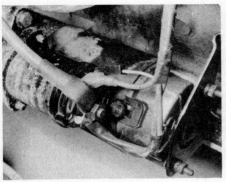

13.9A Starter connections

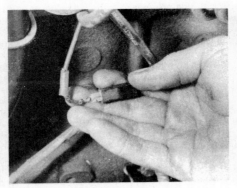

13.9B Coil LT lead connector

13.9C Alternator rear plug, clip and connections

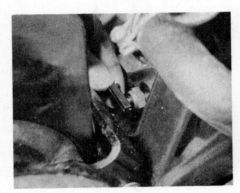

13.9D Oil pressure switch

13.11 Brake servo hose at manifold

12 Disconnect the crankcase breather hose and the distributor vacuum pipe from the intake manifold.
13 Disconnect the carburettor controls, unbolt the carburettor and rest it to one side of the engine compartment.
14 Remove the carburettor heat shield, the throttle bracket and insulating block assembly.
15 Unbolt the clutch slave cylinder and pull it off the pushrod. There is no need to disconnect the hydraulic system, just tie the cylinder up out of the way.
16 Disconnect the speedometer cable from the transmission by unscrewing the knurled ring.

17 Disconnect the exhaust pipes from the manifold.
18 Where fitted, disconnect the engine oil cooler hoses and plug their open ends.
19 Raise the front of the car and support it securely on axle stands so that the following work can be carried out.
20 If the car is equipped with power steering, release the clips which hold the fluid pipes to the transmission and left-hand front engine mounting bracket.
21 Disconnect the exhaust downpipe clip from the gearbox casing.
22 Disconnect the gearchange linkage. To do this, drive out the roll pin which holds the extension rod the to selector shaft. Detach the

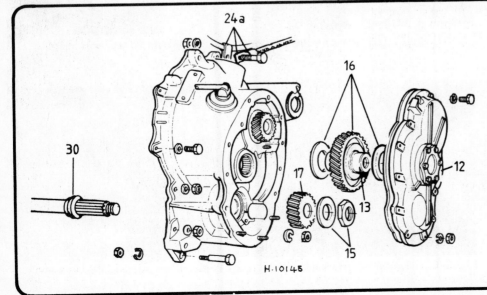

Fig. 1.9 Manual transmission primary drive components (Sec 13)

12 Cover
13 Idler gear bush
15 Input shaft nut and lockplate
16 Idler gear and thrust washers
17 Input shaft gear
24a Earth cable and wiring clip
30 Gearbox input shaft and special washer

H·10145

steady rod from the differential housing and then tie the gearchange rods loosely to the exhaust pipe. Disconnect the reverse lamp switch leads.

23 Drain the engine oil into a suitable container, then refit and tighten the drain plug.

24 Raise the front of the car and support it securely on axle stands placed under the side members.

25 Disconnect the driveshafts from the final drive as described in Chapter 8.

26 Tie the exhaust downpipes to the body crossmember.

27 Release the power steering fluid outlet pipe union on the pump and slacken the pump mounting so that the drivebelt can be withdrawn. Unbolt the pump, move it downwards and tie it to the towing bracket at the front of the engine compartment.

28 Disconnect the bonnet struts from their brackets on the body and tie the bonnet back far enough to permit the engine hoist to be separated when the engine/transmission is removed.

29 Fit lifting brackets to the four cylinder head bolts which are located at the corners of the head.

30 Attach suitable lifting gear and just take the weight of the engine/transmission.

31 Extract the two screws which hold the engine right-hand front mounting bracket to the flywheel housing. Slacken the long bolt which goes through the flexible mounting cushion and swivel the bracket to one side (photo).

32 Remove the long bolt from the left-hand front engine mounting. Remove the mounting bracket, which is held to the cylinder block by two screws and to the transmission casing by two bolts (photo).

33 Remove the long bolts from the engine rear mountings. The engine will now be hanging free, ready for removal from the engine compartment (photos).

34 Hoist the engine/transmission out of the engine compartment and lower it onto the floor.

35 With the engine and transmission removed from the car and supported safely on a bench or held upright on the floor, remove the ten screws and three nuts which secure the transmission primary drive cover and withdraw the cover and joint washer. Be prepared to catch some oil which will be released.

36 Flatten the locking tab and unscrew and remove the input shaft nut. Jam the starter ring gear to prevent the input shaft turning.

37 Remove the idler gear, marking it carefully to ensure that it is refitted the correct way round.

38 Withdraw the gear from the input shaft.

39 Remove the right-hand rear engine mounting bracket (photo).

40 Remove the flywheel housing from the adaptor plate. This necessitates removal of one dowel bolt and seven nuts. Note the nut within the circular recess also the nut pointing in the reverse direction at the lower edge of the housing.

41 Withdraw the special washer from the input shaft.

42 Unbolt and remove the clutch assembly from the flywheel.

43 Flatten the locking tabs and unbolt and remove the flywheel from the crankshaft. Again jam the flywheel ring gear to prevent the flywheel turning when unscrewing the retaining bolts.

44 Remove the adaptor plate screws and withdraw the adaptor plate and joint washer.

45 Extract the four laygear thrust springs from their holes in the gearbox casing.

46 Withdraw the engine oil dipstick and then remove the bolts, screws and nuts which secure the gearbox casing to the engine cylinder block. These consist of six nuts on the oil filter side, seven bolts on the dipstick side, two $\frac{5}{16}$ UNC bolts at the bottom of the front casing. Two bolts are located at the rear, one screwing into the rear main bearing cap and the other one into the cylinder block. Note that the longer screw is on the dipstick side.

47 Lift the engine from the transmission and remove the gasket and O-ring from the oil feed hole.

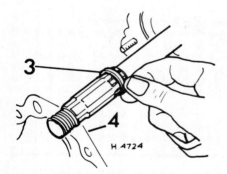

Fig. 1.10 Location of manual transmission input shaft special washer (3) and flywheel housing (4) (Sec 13)

13.31 Engine front right-hand mounting bracket

13.32 Engine front left-hand mounting

13.33A Engine rear right-hand mounting

13.33B Engine rear left-hand mounting

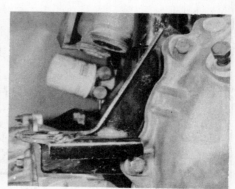

13.39 Right-hand rear mounting bracket

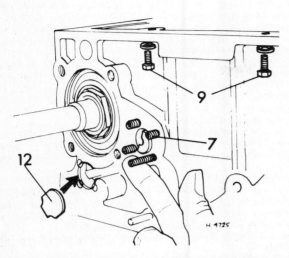

Fig. 1.11 Manual transmission laygear thrust springs (7) engine connecting bolts (9) and selector rod retaining plate (12) (Sec 13)

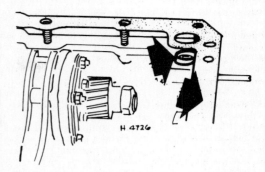

Fig. 1.12 Location of gasket and oil feed O-ring seal, engine separated from manual transmission (Sec 13)

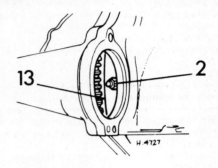

Fig. 1.13 Driveplate (13) to torque convertor connecting bolts (2) (Sec 14)

14 Engine/automatic transmission – removal and separation

1   The operations are very similar to those described in the preceding Section for manual transmission cars but the following differences must be noted.
2   Before removing the carburettor from the manifold, disconnect the kickdown cable and pull it out from behind the suspension pipes.
3   Disconnect the leads from the inhibitor switch, release the retaining clip and pull the outer cable clear of the selector housing.
4   Drain the automatic transmission fluid and refit the drain plug.
5   With the engine and transmission removed from the car and supported safely on a bench or held upright on the floor, work through the starter motor aperture and unscrew each of the four screws which secure the driveplate to the torque converter. The crankshaft will have to be turned by means of its pulley bolt to bring each driveplate bolt in turn into view (Fig. 1.13).
6   Remove the engine rear mounting bracket (two bolts).
7   Remove the harness clip from the top of the torque converter housing.
8   Extract the screw which secures the main transmission casing to the engine crankcase lower extension and is accessible from between the casing and the torque converter housing.
9   Remove the screws which secure the transmission casing to the crankcase, seven screws on the distributor side and six screws on the differential side.
10 Working through the starter motor aperture, push the torque converter fully into its housing, support the engine and lift it from the transmission in a straight line until the driveplate has cleared the converter housing.
11 Detach the transmission to lower extension gasket, unbolt the driveplate from the crankshaft, also the adaptor plate with integral oil seal, and remove the gasket (Fig. 1.15).

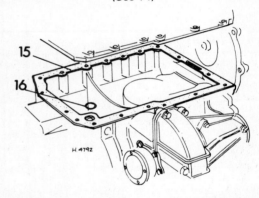

Fig. 1.14 Location of gasket (15) and oil feed O-ring seal (16) – engine separated from automatic transmission (Sec 14)

15 Engine – dismantling general

1   It is best to mount the engine on a dismantling stand, but if this is not available, stand the engine on a strong bench at a comfortable working height. Failing this, it will have to be stripped down on the floor, but at least cover it with hardboard.
2   During the dismantling process, the greatest care should be taken to keep the exposed parts free from dirt. As an aid to achieving this thoroughly clean down the outside of the engine, first removing all traces of oil and congealed dirt.
3   A good grease solvent will make the job much easier, for, after the solvent has been applied and allowed to stand for a time, a vigorous jet of water will wash off the solvent and grease with it. If the dirt is thick and deeply embedded, work the solvent into it with a strong stiff brush.

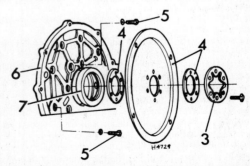

Fig. 1.15 Automatic transmission driveplate components (Sec 14)

3   Lockplate
4   Driveplate and backing plate
5   Screws
6   Adaptor plate
7   Oil seal

4   Finally, wipe down the exterior of the engine with a rag and only then, when it is quite clean, should the dismantling process begin. As the engine is stripped, clean each part in a bath of paraffin.

5   Never immerse parts with oilways in paraffin (eg crankshaft and camshaft). To clean these parts, wipe down carefully with a paraffin dampened rag. Oilways can be cleaned out with wire. If an air line is available, all parts can be blown dry and the oilways blown through as an added precaution.

6   Re-use of old gaskets is false economy. To avoid the possibility of trouble after the engine has been reassembled **always** use new gaskets throughout.

7   Do not throw away old gaskets, for sometimes it happens that an immediate replacement cannot be found and the old gasket is then very useful as a template. Hang up the gaskets as they are removed.

8   To strip the engine, it is best to work from the top down. When the stage is reached where the crankshaft must be removed, the engine can be turned on its side and all other work carried out with it in this position.

9   Wherever possible, refit nuts, bolts and washers finger tight from wherever they were removed. This helps to avoid loss and muddle. If they cannot be fitted then arrange them in a sequence that ensures correct reassembly.

10   Make sure that you have a valve grinding tool and a valve spring compressor.

16 Engine ancillary components – removal

1   Remove the components before engine dismantling commences after reference to the Chapters indicated.

Alternator	*Chapter 10*
Distributor	*Chapter 4*
Starter motor	*Chapter 10*
Carburettor	*Chapter 3*
Power steering pump	*Chapter 11*

17 Engine – complete dismantling

1   With the engine removed, separated from the transmission and clean, commence dismantling for complete overhaul in the following sequence.

2   Remove the crankcase ventilation oil separator, and the oil filler pipe.

3   Unbolt and remove the intake manifold.

4   Unbolt and remove the exhaust manifold noting how the hot air collector backplate fits under the two bolts nearest the crankshaft pulley.

5   Unbolt and remove the camshaft rear cover and discard the gasket.

6   Remove the timing belt cover, remove the belt tensioner pulley and the timing belt.

7   Unbolt and remove the camshaft cover noting the belt cover bracket held by the front two bolts.

8   Remove the camshaft.

9   Take out the shims and tappets (cam followers) and retain them in their originally fitted order.

10   Remove the cylinder head and discard the gasket (Section 8).

11   Jam the starter ring gear and unscrew the crankshaft pulley bolt.

12   Remove the crankshaft pulley, the timing disc, the belt sprocket and the belt guide plate. Remove the Woodruff key.

13   Remove the LED sensor bracket.

14   Remove the oil pump (Section 9).

15   Lay the cylinder block on its side and check for identifying markings on the big-end bearing caps. If they are not marked, dot punch them and an adjacent point on the connecting rod. Number them (1) from the crankshaft pulley end of the engine.

16   Unscrew and remove the big-end nuts, withdraw the big-end caps and their bearing shells. If the shells are to be used again, tape them to their respective big-end caps with masking tape.

17   Scrape the carbon or wear ring from the top of each cylinder bore and then push the piston/connecting rod assembly out of the top of the cylinder block.

18   Retain the bearing shells with their respective connecting rods if new shells are not to be fitted.

19   Piston rings can be removed by sliding two or three feeler blades behind the top ring at equidistant points and sliding the ring off the top of the piston using a twisting motion. Repeat the operation on the other rings, always withdrawing them from the top of the piston.

20   If new rings are to be fitted to the old pistons, thoroughly clean out the piston ring grooves with a piece of broken ring. Remove the hard glaze from the cylinder bores using a piece of fine glass paper and leaving a cross-hatch finish.

21   Removal of the piston from the connecting rod is a job for your dealer as the gudgeon pin can only be removed and refitted using a press and special guide tools.

22   Check that the main bearing caps are numbered. If not, dot punch them 1 to 5, numbering from the crankshaft pulley end. Directional arrows and numbers are usually found on the caps, arrows towards timing end.

23   Unscrew each of the main bearing cap bolts evenly a turn or two. Remove the intermediate and then the front and rear bearing caps.

24   Extract the four small cork joint seals (early models only).

25   Remove the centre main bearing cap and the lower halves of the thrust washers. The main bearing caps are dowelled.

26   Keep the main bearing shells with their respective caps if they are not being renewed.

27   Lift the crankshaft from the crankcase. Mark the location of the crankcase bearing shells if they are not to be renewed. Extract the bottom halves of the crankshaft thrust washers.

18 Cylinder head – dismantling and decarbonising

1   With the cylinder head removed, use a blunt scraper to remove all trace of carbon and deposits from the combustion spaces and ports. A wire brush in an electric drill will speed up the carbon removal operation. Scrape the cylinder head free from scale or old pieces of gasket or jointing compound. Remember that the cylinder head is made of light alloy and the carbon removal operations should be carried out with care.

2   Clean the cylinder head by washing it in paraffin and take particular care to pull a piece of rag through the ports and cylinder head bolt holes. Any dirt remaining in these recesses may well drop onto the gasket or cylinder block mating surface as the cylinder head is lowered into position and could lead to a gasket leak after reassembly is complete.

3   With the cylinder head clean, test for distortion if a history of coolant leakage has been apparent. Carry out this test using a straight edge and feeler gauges or a piece of plate glass. If the surface shows any warping in excess of 0.039 in (0.1015 mm) then the cylinder head will have to be resurfaced which is a job for a specialist engineering company.

Valves

4   Unbolt and remove the camshaft sprocket, the camshaft rear end cover and gasket.

5   Release the camshaft cover bolts evenly and in side-to-side sequence until the pressure of the valve springs on the cam cover bearings can be felt to be relieved.

6   Release the filler cap and remove the cover bracket. Lift the camshaft cover off the cylinder head.

7   Lift the camshaft complete with front oil seal from the cylinder head. Extract the oil seal.

8   Remove each tappet and shim keeping them in their exact fitted order using a box with divisions or other suitable storage device.

9   Using a valve spring compressor, compress each valve in turn, extract the split collets, remove the compressor and then take off the valve spring cap, the spring and the seat. Withdraw the valve. keep all the valves with their associated components in strict order so that they can be refitted in their original positions. Discard the oil seals.

10   Examine the heads of the valves for pitting and burning, especially the heads of the exhaust valves. The valve seatings should be examined at the same time. If the pitting on valve and seat is very slight the marks can be removed by grinding the seats and valves together with coarse, and then fine, valve grinding paste.

11   Where bad pitting has occurred to the valve seats it will be necessary to recut them and fit new valves. If the valve seats are so worn that they cannot be recut, then it will be necessary to fit new valve seat inserts. These latter two jobs should be entrusted to the

local dealer or engineering works. In practice it is very seldom that the seats are so badly worn for refitting, and the owner can easily purchase a new set of valves and match them to the seats by valve grinding.

12 Valve grinding is carried out as follows: Smear a trace of coarse carborundum paste on the seat face and apply a suction grinder tool to the valve head. With a semi-rotaty motion, grind the valve head to its seat, lifting the valve occasionally to redistribute the grinding paste. When a dull matt even surface finish is produced on both the valve seat and the valve, wipe off the paste and repeat the process with fine carborundum paste, lifting and turning the valve to redistribute the paste as before. A light spring placed under the valve head will greatly ease this operation. When a smooth unbroken ring of light grey matt finish is produced, on both valve and valve seat faces, the grinding operation is completed.

13 Scrape away all carbon from the valve head and the valve stem. Carefully clean away every trace of grinding compound, taking great care to leave none in the ports or in the valve guides. Clean the valves and valve seats with a paraffin soaked rag, then with a clean rag, and finally if an air line is available, blow the valves, valve guides and valve ports clean.

14 After a considerable mileage, the valve guide bores may wear oval in shape. Insert each valve into its respective guide and check for excessive clearance. Establish whether it is the guide or valve stem which is worn or both, by reference to the Specifications or by testing a new valve in the guide.

15 Renewal of the valve guides is best left to your dealer but for those owners who have access to a small press the following information is given.

16 Heat the cylinder head uniformly in boiling water and then press the old guide from the head so that it emerges from the cylinder block mating face.

17 Keep the cylinder head up to temperature (212°F/100°C) and press the new guide in from the top face of the cylinder head until it is projecting 0.394 in (10.0 mm). A distance piece must be made up to prevent the valve guide being pressed in too far. A positive interference fit of the guide in the head is essential.

18 Examine the valve springs. If their free or fitted lengths do not match up with those specified or if they have been in use for 30 000 miles (48 000 km) or more, then they should be renewed.

Reassembly and refitting

19 Fit new valve stem oil seals, Smear the stem of the first valve in oil and insert it into its guide. Fit the spring seat, the valve spring, the spring cap and then compress the spring using the compressor. Insert the split collets and then gently release the compressor in order not to displace the collets (photo).

20 Refit the remaining valves in a similar way. When they are all fitted, tap the end of each valve stem using a plastic-faced hammer or a hammer with a block of hardwood interposed. This will settle the valve components ready for checking the valve clearances.

21 Place the valve adjusting shims in their original valve spring cap recesses. If the valves have been ground in more than very lightly it will usually be necessary to reduce each shim thickness by 0.010 in (0.25 mm) and new shims should be obtained to achieve this or the original ones interchanged as described in Section 11. Fit the tappets (cam followers) photos.

22 Oil the camshaft journals and tappets and locate the camshaft on the cylinder head. Adjust the valve clearances as described in Section 11.

23 Fit the camshaft, oil seal and camshaft cover as described in Section 5.

Piston crowns

24 An essential part of decarbonising is to remove carbon from the piston crowns.

25 If the work is being carried out by removing the cylinder head with the engine still in the car then proceed in the following way.

26 When removing carbon from the pistons and upper parts of the cylinder bores, it is essential that care is taken to ensure that no carbon gets into the cylinder bores as this could scratch the cylinder walls or cause damage to the piston and rings. To ensure this does not happen, first turn the crankshaft so that two of the pistons are at the top of

18.19A Inserting valve into guide

18.19B Valve spring

18.19C Valve spring cap

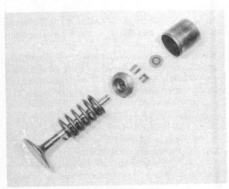
18.19D Valve components

18.21A Valve adjusting shim

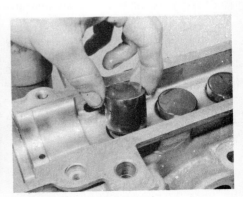
18.21B Tappet (cam follower)

their bores. Stuff rag into the other bores or seal them off with paper and masking tape. Tape over the oilways and waterways to prevent particles of carbon entering the cooling system and damaging the coolant pump.

27 Rotate the crankshaft and repeat the carbon removal operations on the remaining pistons and cylinder bores.

28 Thoroughly clean all particles of carbon from the bores and then inject a little light oil round the edges of the pistons to lubricate the piston rings.

29 Refit the cylinder head as described in Section 8.

19 Examination and renovation – general

1 Most dismantling and overhaul work will be undertaken to rectify a known fault or to renew an obviously faulty component.

2 Where an engine has covered a high mileage, however, and is worn generally, then after dismantling, the components should be examined as suggested in this Section.

3 Where the majority of parts are found to be worn excessively and have tolerances outside those specified then it will probably be more economical to exchange the engine for a fully reconditioned one.

20 Examination and renovation of engine components

Crankcase

1 Examine the webs and areas around bolt holes for cracking. Clean out oil passages with a piece of wire.

2 It is possible to rectify stripped threads by fitting special inserts, but this is really a job for your local motor engineering works.

Crankshaft and main bearings

3 Examine the crankpin and main journal surfaces for signs of scoring or scratches. Check the ovality of the crankpins at different positions with a micrometer. If more than specified out of round, the crankpin will have to be reground. It will also have to be reground if there are any scores or scratches present. Also check the journals in the same fashion.

4 If it is necessary to regrind the crankshaft and fit new bearings your local Austin-Rover garage or engineering works will be able to decide how much metal to grind off and the size of new bearing shells.

5 Always check the spigot bush (located in the crankshaft rear flange) and renew it if necessary.

6 On cars with manual transmission, extract the bush by tapping a thread into it. Screw in a bolt and by using a distance piece, withdraw the bush. When driving in the new bush, make sure that it is inserted until dimension (A) Fig. 1.16 is as specified.

7 On cars with automatic transmission, remove the bush as described in the preceding paragraph.

8 Refer to Fig. 1.17. Measure the depth (A) of the bush recess from the machined face of the crankshaft flange and then fit a bush, selected after reference to the following table, which will have a narrow bore length (B).

9 The bush if correctly selected and fitted will then provide a driveplate pre-load depth (C) of between 0.030 and 0.051 in (0.76 and 1.29 mm).

Recess depth (A)	Bush to fit (length of narrow bore B)
0.969 to 0.972 in (24.61 to 24.86 mm)	0.930 to 0.935 in (23.62 to 23.74 mm)
0.980 to 0.996 in (24.89 to 25.29 mm)	0.945 to 0.950 in (24.00 to 24.13 mm)
0.997 to 1.013 in (25.32 to 25.73 mm)	0.962 to 0.967 in 24.43 to 24.56 mm)
1.014 to 1.026 in (25.79 to 25.06 mm)	0.978 to 0.984 in (24.68 to 24.99 mm)

10 Drive the selected bush into the full depth of the crankshaft recess.

Connecting rods and big-end bearings

11 Big-end bearing failure is indicated by a knocking from within the crankcase and is accompanied by a slight drop in oil pressure. Examine the big-end bearing surfaces for pitting and scoring. Renew the shells

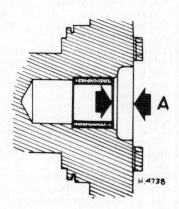

Fig. 1.16 Crankshaft spigot bush fitting diagram (manual transmission) (Sec 20)

A   0.30 in (7.62 mm)

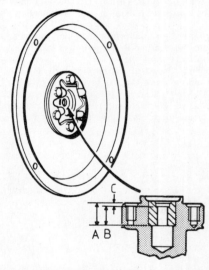

Fig. 1.17 Crankshaft spigot bush fitting diagram (automatic transmission) (Sec 20)

For values of A, B and C refer to text

with ones of the same original size, or if the crankshaft has been reground, the correct undersize big-end bearing shells will be supplied by the repairer.

12 The gudgeon pin is an interference fit in the connecting rod small end and a push fit in the piston. It is recommended that where the connecting rod must be renewed due to twist or to slackness in the small-end, then this operation is left to your dealer due to the need for special tools.

Cylinder bores

13 The cylinder bores must be examined for taper, ovality, scoring and scratches. Start by carefully examining the top of the cylinder bores. If they are at all worn a very slight ridge will be found on the thrust side. This marks the top of the piston ring travel. The owner will have a good indication of the bore wear prior to dismantling the engine, or removing the cylinder head. Excessive oil consumption accompanied by blue smoke from the exhaust is a sure sign of worn cylinder bores and piston rings.

14 Another way to check for bore wear or valve problems is to use a compression tester before dismantling an engine. Compression testers are screwed in place of one of the spark plugs. Disconnect the coil HT lead to prevent the engine firing and then depress the accelerator fully then spin the engine on the starter motor. Record the reading on the tester and then repeat the operation on the remaining cylinders.

15 If the compression readings do not approximate the figure specified in the Specifications then assume the bores are worn. If only one reading is low, suspect a valve not seating correctly.

16 Measure the bore diameter just under the ridge with a micrometer and compare it with the diameter at the bottom of the bore which is not subject to wear. If the difference between the two measurements is more than 0.008 in (0.2032 mm) then it will be necessary to fit special pistons and rings or to have the cylinders rebored and fit oversize pistons. If no micrometer is available remove the rings from a piston and place the piston in each bore in turn about $\frac{3}{4}$ in below the top of the bore. If an 0.0012 in (0.0254 mm) feeler gauge slid between the piston and the cylinder wall requires less than a pull of between 1.1 and 3.3 lbf (0.5 and 1.5 kgf) to withdraw it, using a spring balance, then remedial action must be taken. Oversize pistons are available.

17 These are accurately machined to just below the indicated measurements so as to provide correct running clearances in bores bored out to the exact oversize dimensions.

18 If the bores are slightly worn but not so badly worn as to justify reboring them, then special oil control rings and pistons can be fitted which will restore compression and stop the engine burning oil. Several different types are available and the manufacturer's instructions concerning their fitting must be followed closely.

19 If new pistons are being fitted and the bores have not been reground, it is essential to slightly roughen the hard glaze on the sides of the bores with fine glass paper so the new piston rings will have a chance to bed in properly.

Pistons and piston rings

20 If the original pistons are being refitted, carefully remove the piston rings (see Section 17).

21 Clean the grooves and rings free from carbon, taking care not to scratch the aluminium surfaces of the pistons.

22 If new rings are to be fitted, then order the top compression ring to be stepped to prevent it impinging on the 'wear ring' which will almost certainly have been formed at the top of the cylinder bore.

23 Before fitting the rings to the pistons, push each ring in turn down to its lowest limit of normal travel in its respective cylinder bore (use an inverted piston to do this and to keep the ring square in the bore) and measure the ring end gap. The gaps should be as listed in the Specifications Section.

24 Now test the side clearance of the compression rings which again should be as shown in the Specifications Section.

25 Where necessary a piston ring which is slightly tight in its groove may be rubbed down holding it perfectly squarely on an oilstone or a sheet of fine emery cloth laid on a piece of plate glass. Excessive tightness can only be rectified by having the grooves machined out.

26 On these engines, there are two compression rings and one oil control ring (Fig. 1.18). Assemble in the following way:

> **Oil control ring,** *located in bottom groove, set the end gaps in rails and expander at 90° to each other*
> **Stepped compression ring,** *word 'TOP' uppermost, located in stepped groove from top of piston*
> **Plain compression ring,** *word 'TOP' uppermost, located in top groove of piston*

Camshaft and bearings

27 Wear in the camshaft bearings can only be rectified by renewal of the cam cover and the cylinder head complete.

28 The camshaft itself should show no signs of wear, but, if very slight scoring on the cams is noticed, the score marks can be removed by gentle rubbing down with a very fine emery cloth. The greatest care should be taken to keep the cam profiles smooth.

29 Examine the skew gear for wear, chipped teeth or other damage.

30 Carefully examine the camshaft end cover. Excessive endfloat

Fig. 1.18 Piston ring fitting diagram (Sec 20)

31 Oil control bottom rail	*36 Second compression ring*
32 Expander	*37 Top compression ring*
34 Oil control top rail	

(more than 0.007 in (0.17 mm)), will be visually self-evident and will require the fitting of a new end cover.

Timing belt and sprockets

31 If the timing belt has stretched beyond the ability of the tensioner pulley to tension it as specified then the belt must be renewed. Also discard the belt if it shows signs of cracking or general deterioration.

32 The sprocket teeth should be examined for wear and chipping and renewed if evident.

33 Always renew the timing belt at the intervals specified in Routine Maintenance whether it appears worn or not.

Tappets and shims

34 Examine the contact face of the tappet and the ground faces of the shim for grooves or pitting and renew any that show such signs. Remember that the shim thicknesses were recorded at dismantling and any replacement must be identical to the one discarded.

Flywheel (driveplate – automatic transmission) and starter ring gear

35 Worn teeth on the starter ring gear will mean removal of the flywheel, on manual transmission cars, or the torque converter, on automatic transmission models, to which the starter ring gear is attached.

36 If the flywheel clutch surface is scored or grooved or if many tiny cracks are to be seen, caused by overheating, a new flywheel is recommended.

37 Examine the bolt holes on all components for cracks or elongation.

Oil pump

38 Unscrew and remove the oil pressure switch and extension hollow pillar bolt.

39 The dismantling operations are only to enable inspection of the internal components to be carried out. If parts are found to be worn then the complete oil pump must be renewed, individual spare parts not being available.

40 Extract the split-pin from the relief valve housing. Force the plug

20.41A Using an Allen key to remove oil pump screws

20.41B Oil pump backplate

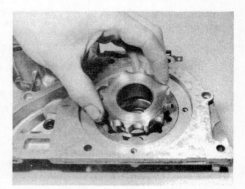
20.42A Oil pump inner rotor

20.42B Oil pump outer rotor

20.43 Checking inner to outer rotor tip clearance

20.44 Checking outer rotor to body clearance

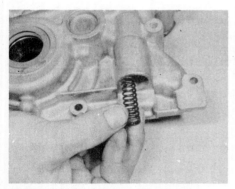
20.47A Oil pump relief valve spring

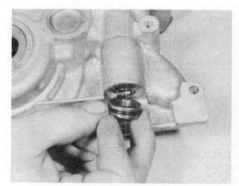
20.47B Oil pump relief valve plunger

21.3 Oil seal directional (shaft rotation) arrow

from the housing by pushing the relief valve spring down with a screwdriver inserted through the oil return hole.
41   Mark the relative position of the backplate to the oil pump housing. Extract the backplate screws and remove the backplate (photos).
42   Withdraw the internal components, clean them and smear with clean engine oil (photos).
43   Refit the outer rotor and check the clearance between it and the pump body (photo).
44   Refit the inner rotor and check the clearance between rotors as shown in the diagram (photo).
45   Using a straight-edge, check the outer ring endfloat. Use feeler blades for all checking purposes. Where the tolerances are outside those specified, renew the oil pump complete.
46   Assemble the pump by fitting the backplate so that the marks made before dismantling are in alignment.
47   Fit a new O-ring seal to the pressure relief valve plug, insert the

plunger and spring into the oil pump housing, depress the components with the plug and insert a new split-pin (photos).

21 Engine reassembly – general

1   To ensure maximum life with reliability from a rebuilt engine, not only must everything be correctly assembled but all components must be spotlessly clean and the correct spring or plain washers used where originally located. Always lubricate bearing and working surfaces with clean engine oil during reassembly of engine parts.
2   Before reassembly commences, renew any bolts or studs the threads of which are damaged or corroded.
3   As well as your normal tool kit, gather together clean rags, oil can, a torque wrench and a complete (overhaul) set of gaskets and oil seals (photo).

22 Engine – reassembly

1   With all components conveniently laid out, stand the cylinder block upside down ready for fitting its internal parts.

Crankshaft

2   Before refitting the crankshaft, carefully clean out the crankshaft oilways with a piece of wire.

3   Fit the bearing shells into their crankcase recesses. If the old shells are being refitted, make sure that they are returned to their original locations. Shells 1, 3 and 5 are grooved, 2 and 4 are plain (photo).

4   Oil the shells and locate the thrust washer half sections (without the tabs) either side of the centre main bearing so that the lubrication grooves are towards the crankshaft (photo).

5   It is important that before the crankshaft is refitted the shell bearing recesses and the mating faces of the bearing caps are wiped absolutely clean. The mating face of the rear main bearing cap should be smeared with jointing compound before fitting and the four front and rear main bearing cap cork seals renewed after first having soaked them in engine oil. Later models do not have cork seals, but a bead of RTV type sealant should be used instead.

6   Lower the crankshaft into position and fit the main bearing caps (dowelled) complete with bearing shells. Make sure that the halves of the centre bearing thrust washers have their tabs engaged in the cap recesses and the oil grooves are facing outwards towards the crankshaft (photos).

7   Refit the main bearing cap bolts and tighten to the specified torque (photo).

8   At this point check the crankshaft endfloat by pushing and pulling the crankshaft and using a dial gauge or feeler blade inserted between the centre main bearing cap and the machined face of the crankshaft web (Fig. 1.19). If it is outside the permitted tolerance, renew the thrust washers which are available in standard or oversizes (photo).

Connecting rods and pistons

9   To refit a piston/connecting rod assembly, first set the end gaps of

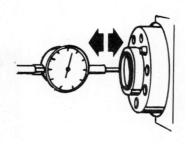

Fig. 1.19 Checking crankshaft endfloat using a dial gauge (Sec 22)

the top and second rings at 90° to each other and away from the thrust side of the piston.

10   Oil the piston rings liberally and fit a piston ring compressor.

11   Smear the cylinder bores with engine oil.

12   Wipe the connecting rod and cap shell bearing seats absolutely clean and fit the shells.

13   Insert the rod into its appropriate bore so that the compressor stands squarely on the top face of the cylinder block with the FRONT or triangular mark on the piston crown towards the crankshaft pulley end of the engine (photo).

14   Drive the piston/rod assembly into the bore by applying the wooden handle of a hammer to the piston crown and striking the head of the hammer. The rings will slide into the bore as the compressor is released (photo).

15   Turn the crankshaft so that the crankpin is at the lowest point of its 'throw' and connect the rod to it. Oil the big-end cap shell and fit the cap so that its number is adjacent to the one on the side of the rod (photos).

16   Screw on and tighten the big-end nuts to the specified torque.

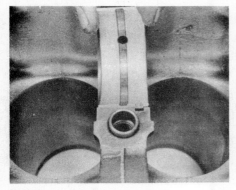

22.3 Grooved main bearing shell

22.4 Centre bearing shell and thrust washer

22.6A Lowering crankshaft into position

22.6B Fitting centre main bearing cap

22.7 Tightening main bearing cap bolt

22.8 Checking crankshaft endfloat

22.13 Piston crown marking

22.14 Piston ring compressor

22.15A Big-end cap and shell

22.15B Fitting big-end cap

22.16 Tightening big-end cap nut

22.19A Coolant pump

22.19B Coolant pump fitted

22.20A Timing belt guide plate

22.20B Timing belt sprocket

Repeat the fitting operations on the remaining three connecting rod/piston assemblies (photo).

Oil pump and adjacent parts
17 Fit the oil pump as described in Section 9, paragraphs 11 to 17.
18 Fit the LED sensor bracket and ignition timing pointer.
19 Fit the coolant pump (Chapter 2) making sure that the captive bolts for the belt tensioner and the spigot bolt for the timing belt cover grommet are in position (photos).
20 To the front end of the crankshaft, fit the Woodruff key, the timing belt guide plate, the belt sprocket, the timing disc and crankshaft pulley. Fit the three pulley bolts (photos).
21 Screw in the pulley bolt using a new lockplate. Tighten the bolt to the specified torque while jamming a crankshaft web against the inside of the crankcase using a block of wood, alternatively, join the teeth of the starter ring gear. Bend up a lockplate tab (photos).

Cylinder head
22 Refit the cylinder head as described in Section 8. The head should be complete with camshaft and shims with the valve clearances adjusted as described in Section 11.
23 Fit the timing belt and tensioner and tension the belt as described in Section 7. Fit the timing belt cover.
24 Fit the camshaft rear cover plate with a new gasket.
25 Using new gaskets, bolt on the manifolds to the cylinder head. Remember that the hot air collector fits under the two lower exhaust manifold bolts nearest the crankshaft pulley (photos).
26 Screw the vacuum pipe connector into the manifold.
27 Fit the oil separator for the crankcase ventilation system and the oil filler pipe with gasket (photos).
28 Fit the coolant pump pulley (photo).
29 Screw on a new oil filter (Section 2).
30 Screw in the spark plugs.

22.20C Timing disc

22.20D Crankshaft pulley

22.21A Crankshaft pulley lockplate

22.21B Method of locking flywheel ring teeth

22.25A Manifold gasket

22.25B Hot air collector backplate on exhaust manifold

22.25C Fitting hot air collector to backplate

22.25D Hot air collector

22.27A Crankcase vent oil separator

22.27B Crankcase vent oil separator bolted to crankcase

22.27C Oil filler pipe bolted to crankcase

22.28 Coolant pump pulley

23 Engine ancillary components – refitting

1   Refitting may be carried out before installing the engine/transmission to the car or after installation.
2   Refitting should be carried out after reference to the following Chapters.

   Alternator – Chapter 10
   Distributor – Chapter 4
   Carburettor – Chapter 3

3   Tension the drivebelt as described in Chapter 2, Section 10.

24 Engine/manual transmission – reconnection and refitting

1   Reconnect the engine to the transmission by referring to Section 13 and reversing operations described in paragraphs 35 to 47. The following essential points must be observed.
2   Locate a new O-ring seal to the oil feed and place new flange gaskets in position and lower the engine onto the transmission casing. Insert and tighten the bolts (photos).
3   Position the layshaft cut-away to mate with the recess in the adaptor plate, then insert the four laygear thrust springs into their holes in the gear case (photo).
4   Check that the gear selector and retaining plate is still in position.
5   Fit a new gasket followed by the adaptor plate. Grease the oil seal lips liberally (photos).
6   Fit the flywheel to the crankshaft flange. Apply thread locking fluid to the bolt-threads, locate a new lockplate and tighten the bolts to the specified torque. Bend up the lockplate tabs (photo).
7   Fit and centralise the clutch as described in Chapter 5.
8   Locate the special washer on the input shaft (photo).
9   Fit the flywheel housing complete with clutch release bearing and

24.2A Oil feed O-ring seal

24.2B Transmission flange gasket

24.2C Lowering engine onto transmission

24.2D Engine/transmission connected

24.3 Layshaft cut-away and thrust springs

24.5A Adaptor plate gasket

24.5B Adaptor plate

24.6A Flywheel

24.6B Applying thread locking fluid to flywheel bolt

24.6C Tightening flywheel bolt

24.6D Flywheel bolt lockplate tabs bent up

24.8 Input shaft special washer

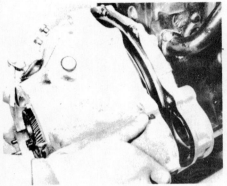
24.9A Fitting flywheel housing

24.9B Flywheel housing recessed nut

24.10A Primary idler gear thrust washer

24.10B Fitting large primary idler gear

24.10C Primary input shaft gear

24.10D Input shaft gear lockplate

24.10E Primary gear nut

24.10F Tightening primary gear nut

24.10G Primary gear nut lockplate bent up

fork. When fitting the connecting bolts, note the location of the recessed nut and always use a new housing flange gasket (photos).
10 Assemble the primary drive by fitting the thrust washer, the large idler gear and the second thrust washer. Fit the remaining gear to the input shaft, followed by the lockplate and screw on the nut. Tighten the nut to the specified torque and bend over the lockplate tab (photo).
11 Locate a new gasket and fit the primary drive cover. Remember to fit the clutch release fork return spring bracket under the cover bolts. Fit the small end cover plate and bend up the lockplate tabs. Prime the primary gear with oil as described in Section 6, Chapter 6. Fit the starter motor (photos).
12 Refit the engine/transmission by reversing the operations described in Section 13, paragraphs 1 to 34, making sure to include the following:
13 Using the hoist, lower the engine carefully into the engine compartment taking care that it does not damage adjacent components (photo).

14 Connect the engine mountings. Remember how the engine mountings are designed: with the rear mountings, the pressed steel brackets are located under the mounting cushions. The left-hand front mounting is also attached in this way, but with the right-hand front mounting, the pressed steel bracket is above the mounting cushion which must have a large steel washer both above and below the cushion. Make sure that the power steering fluid pipes are inboard of the left-hand front mounting.
15 Insert the driveshafts into the differential, making sure that they engage positively with their retaining rings (refer to Chapter 7) (photo).
16 Reconnect the suspension swivel joints and the tie-rod end ball-joints.
17 Connect the exhaust pipes to the manifold, using new asbestos seals or sealing compound.
18 Fit the power steering pump and tension the drivebelt.
19 Reconnect the gearchange rod and steady rod (photo).
20 Reconnect the oil cooler pipes (where fitted).

24.11A Primary gear cover gasket

24.11B Fitting primary gear cover

24.11C Clutch release fork return spring

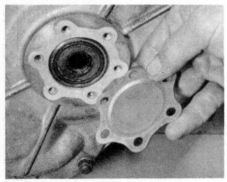
24.11D End cover plate and gasket

24.11E End cover lockplate

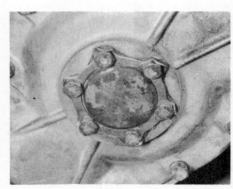
24.11F End cover lockplate tabs bent up

24.13 Engine/transmission ready for installation

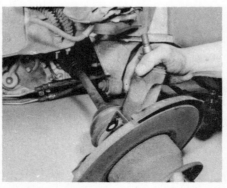
24.15 Reconnecting driveshafts and upper swivel

24.19 Gearchange and steady rods

21 Reconnect the speedometer cable to the transmission.
22 Bolt on the clutch slave cylinder and attach the return spring.
23 Fit the heat shield, the insulator block and the carburettor.
24 Reconnect the throttle and choke cables.
25 Reconnect the crankcase breather hose.
26 Connect the brake servo and distributor vacuum hoses to the intake manifold.
27 Fit the air cleaner and the air intake hose assembly.
28 Connect all the electrical leads including the earth straps (photo).
29 Fit the battery carrier and the horns and leads.
30 Fit the radiator and expansion tank and reconnect the hoses.
31 Connect the heater hose to the cylinder block.
32 Reconnect the battery.
33 Fill the engine with oil and coolant and fill the power steering system with specified fluid.

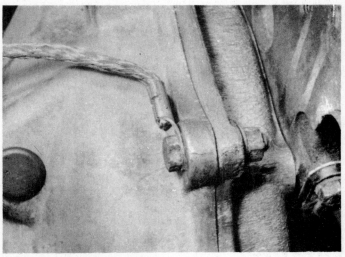

24.28 Flywheel housing earth strap

25 Engine/automatic transmission – reconnection and refitting

1 The operations are very similar to those described in the preceding Section, but remember to reconnect the downshift cable to the carburettor, also reconnect the leads to the inhibitor switch.
2 Fill the engine with oil and coolant, fill the automatic transmission and the power steering system.

26 Initial start-up after major overhaul

1 Adjust the throttle stop screw to provide a faster than normal idling speed. This is to offset the stiffness of new components.
2 Start the engine and check for oil and water leaks.
3 After the first 1000 miles (1600 km) it is recommended that the cylinder head bolts and nuts are checked for tightness with the engine **cold**. To do this release the first bolt in the tightening sequence one

quarter of a turn and then tighten to the specified torque. Repeat the operations on all the remaining bolts in sequence (see Fig. 1.7). Torque tightening of the cylinder head bolts is not required after the first 1000 miles (1600 km) with a new car as the bolts are tightened during production using a new technique which does away with the need for this.
4 Depending upon the number of new internal components which have been fitted, the engine speed should be restricted for the first few hundred miles. It is recommended that the engine oil and filter are changed after the initial 1000 miles (1600 km) running.

27 Fault diagnosis – engine

Symptom	Reason(s)
Engine will not turn over when starter switch is operated	Flat battery Bad battery connections Bad connections at solenoid switch and/or starter motor Defective starter motor
Engine turns over normally but fails to start	No spark at plugs No fuel reaching engine Too much fuel reaching the engine (flooding)
Engine starts but runs unevenly and misfires	Ignition and/or fuel system faults Incorrect valve clearances Burnt out valves Worn out piston rings
Lack of power	Ignition and/or fuel system faults Incorrect valve clearances Burnt out valves Worn out piston rings
Excessive oil consumption	Oil leaks from crankshaft, camshaft, cover gasket, oil filter gasket, sump plug washer Worn piston rings or cylinder bores resulting in oil being burnt by engine Worn valve guides and/or defective inlet valve stem seals
Excessive mechanical noises from engine	Wrong valve clearances Worn crankshaft bearings Worn cylinders (piston slap)

Note: *When investigating starting and uneven running faults do not be tempted into snap diagnosis. Start from the beginning of the check procedure and follow it through. It will take less time in the long run. Poor performance from an engine in terms of power and economy is not normally diagnosed quickly. In any event the ignition and fuel systems must be checked first before assuming any further investigation needs to be made.*

Chapter 2
Cooling, heating and ventilation systems

Contents

Specifications

System type .... Pressurised with expansion tank. Belt-driven coolant pump and electric cooling fan

Pressure cap rating .... 15 lbf/in² (1.04 bar)

Thermostat .... 180°F (82°C)

Fan thermostatic switch (circuit closing temperature) .... 194°F (90°C)

Coolant
Type/specification .... Antifreeze to BS 3151, 3152 or 6580 (Duckhams Universal Antifreeze and Summer Coolant)
Capacity .... 12.25 Imp pts (6.95 litres)

Torque wrench settings	lbf ft	Nm
Cylinder block drain plug (earlier models only)	27	37
Coolant pump mounting bolts	8	11
Coolant pump pulley bolts	9	12
Thermostat housing/filler fixing screw	8	11
Coolant temperature switch	5	7

1   General description and maintenance

1   The cooling system consists of a front mounted radiator, a coolant pump, belt-driven from the crankshaft pulley, and the necessary interconnecting hoses.
2   The system is pressurized and incorporates an expansion tank which removes the necessity of regular topping-up.
3   The expansion tank which is connected to the top of the radiator accepts the overflow of coolant as the cooling system warms up.

When the system cools, the pressure in the radiator drops and the displaced coolant returns from the expansion tank.
4   When the car is in normal forward motion, the ram effect of the air cools the coolant in the radiator tubes but when the car is stationary or moving slowly in hot weather, a supplementary electrically-operated fan cuts in by means of a thermostatically controlled switch located on the side of the radiator. A thermostat is located in the coolant filler neck on the cylinder head to prevent circulation of coolant until the engine has warmed up after starting from cold.
5   Coolant should rarely be required to top up the expansion tank to

its level mark. If it is, check for system leaks. Keep the alternator/coolant pump drivebelt in good condition and correctly tensioned.

2 Cooling system – draining

1 Unscrew and remove the pressure cap from the expansion tank. Unscrew it very slowly if the system is hot in order that the internal pressure may be released.
2 On earlier models, remove the cylinder block drain plug and deflect the coolant away from the starter motor (using a piece of cardboard) into a suitable container. Squeezing the hose between the coolant pump and the metal pipe will assist in ejecting the coolant from the drain plug (later models do not have a block drain plug) (photo).
3 Release the clip and disconnect the water pump hose from the coolant filler neck.
4 Disconnect the radiator bottom hose and allow any remaining coolant to drain.
5 On later models built from November 1982 a cylinder block drain plug is not fitted and draining should be carried out just by disconnecting the radiator bottom hose.
6 It should be noted that draining the cooling system will automatically drain the expansion tank.

3 Cooling system – flushing

1 Place the heater control to HOT and leave the radiator bottom hose disconnected and the cylinder block drain plug out (earlier models only).
2 Remove the filler cap and insert a cold water hose. Allow the water to run until it emerges from both outlets quite clean.
3 In severe cases of contamination, remove the radiator (Section 6) and invert it. Place the cold water hose in the bottom outlet (which is now at the top) and reverse flush it.
4 If the cooling system is blocked or badly corroded due to neglect in routine changing of the antifreeze mixture or failure to maintain its strength, then a chemical cleaner or descaler can be used strictly in accordance with the manufacturer's instructions.

4 Cooling system – filling

1 Reconnect the coolant hoses and screw in and tighten the cylinder block drain plug (earlier models only).

2 Remove the cap from the coolant filler neck on the cylinder head and extract the thermostat from the filler neck.
3 Slowly pour coolant into the filler neck until it can be seen to flow into the expansion tank. Once the expansion tank is half full, refit the expansion tank cap (photo).
4 For the remainder of the operation, do not remove the expansion tank cap unless the coolant filler cap on the cylinder head is in position.
5 Fill the filler neck on the cylinder head brim full and fit the cap.
6 Remove the expansion tank cap and start and run the engine at a fast idle for 30 seconds, at the same time compressing the top hose with the fingers a number of times.
7 Switch off the engine, check that the expansion tank is half full and refit the cap to it.
8 Now remove the coolant filler cap on the cylinder head. Check that the rubber O-ring seal is in good condition and fit the thermostat.
9 Top up to the brim of the filler neck and fit the cap.

5 Antifreeze mixture

1 The cooling system should always be filled with antifreeze mixture of suitable strength as apart from protecting the engine from frost damage, it will also prevent corrosion in the system.
2 It is recommended that the system is drained and refilled with fresh mixture every two years.
3 Before adding the antifreeze mixture, check the tightness of all hose clips as the mixture is more searching than plain water and will find the smallest leak in the system.
4 The quantity of antifreeze which should be used for various levels of protection according to climate, is given in the following table:
5 To maintain adequate protection against corrosion, a minimum concentration of $33\frac{1}{3}\%$ antifreeze is recommended.

Concentration	Quantity of antifreeze	Protects to
$33\frac{1}{3}\%$	3.5 Imp pts (2.0 litres)	-2°F (-19°C)
50%	5.25 Imp pts (3.0 litres)	-33°F (-36°C)

6 Any topping-up of the antifreeze should be carried out using a solution made up in similar proportions to the original, in order to avoid dilution.
7 When climatic conditions do not require the use of antifreeze, a corrosion inhibitor should be used instead of plain water, in order to protect the aluminium content of the engine.

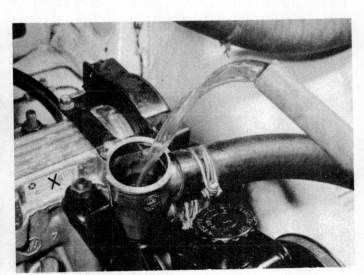

2.2 Cylinder block drain plug (earlier models)

4.3 Filling cooling system

6.2 Radiator bottom hose

6 Radiator – removal, inspection and refitting

1   Drain the cooling system as described in Section 2.
2   Disconnect the radiator top hose from the coolant filler neck on the cylinder head and the bottom hose from the radiator (photo).
3   Disconnect the expansion tank hose from the radiator (photo).
4   Remove the two bolts which hold the radiator top securing brackets to the crossmember (photo).

5   Tilt the top of the radiator towards the rear of the car and disconnect the electric fan wiring plug and the thermostatic switch wiring plug.
6   Lift the radiator from the engine compartment. The metal coolant pipe below the radiator is held in position by hose clips around tongues. The pipe has a cover over it (photos).
7   If the radiator has been removed because of a leak, it should be professionally repaired owing to the need to localise the heat, otherwise further damage will be caused. Better still, exchange it for a reconditioned one.
8   Whenever the radiator is removed, take the opportunity of brushing away all accumulations of flies or dirt from the radiator fins or by applying air from a compressed air line in the reverse direction to normal airflow.
9   It is a good plan to have the expansion tank pressure cap tested at a service station periodically and if faulty, renewed with one of similar pressure rating.
10 Refitting is a reversal of removal. Fill the system, as described in Section 4 of this Chapter.

7 Radiator fan – removal and refitting

1   Disconnect the electrical leads from the thermostatic fan control switch.
2   Disconnect the electrical leads from the radiator fan motor.
3   Unclip (not disconnect) the expansion hose from the radiator.
4   Release the radiator brackets by extracting the fixing screws.
5   Remove the four fan cowl screws from the radiator.
6   Lift the radiator from its base mounting bushes, raise the top hose and withdraw the fan assembly from between the radiator and the front panel. The radiator will have to be moved towards the engine in order to provide sufficient space for the fan assembly to pass.
7   Refitting is a reversal of removal.
8   Automatic transmission models have twin cooling fans. The removal procedure is similar to that just described.

6.3 Expansion tank hose

6.4 Radiator top mounting

6.6A Removing radiator

6.6B Radiator lower coolant pipe

6.6C Radiator lower coolant pipe to hose connection

6.6D Radiator lower coolant pipe cover

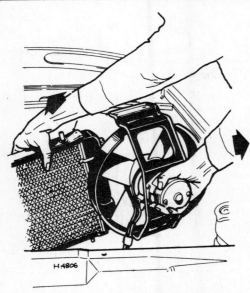

Fig. 2.1 Removing radiator fan assembly (Sec 7)

3   Remove the three screws which secure the fan motor to the cowl and separate the two components.
4   Cut through the plastic weatherproof cover which seals the motor end cover.
5   Remove the two motor tie-bolts and then withdraw the end cover complete with armature from the motor yoke, noting the assembly marks on the end cover and on the yoke.
6   Remove the circlip from the armature spindle and withdraw the two shim washers and wave washer.
7   Withdraw the armature assembly from the end cover.
8   Withdraw the thrust washer from the armature spindle and then extract the circlip from the armature spindle.
9   Remove the three screws to release the brush carrier assembly from the end cover.
10 Renew any worn components particularly the brushes after comparison with new ones.
11 Clean the commutator with a fuel moistened cloth and if essential, polish it with fine glass paper, not emery cloth. Do not undercut the segment insulators.
12 Reassembly is a reversal of dismantling but observe the following points: Lubricate sparingly, the bearings and bushes. Make sure that the yoke and end cover mating marks are in alignment. Position the wave washer **between** the two shims.
13 When fitting the fan to the shaft, ensure that the boss of the fan is away from the motor and engage the grub screw with the first indent in the shaft.

8   Radiator fan motor – overhaul

1   With the radiator fan assembly removed as described in the preceding Section, carry out the following operations.
2   Slacken the grub screw which secures the fan hub to the motor spindle and withdraw the fan blades.

9   Thermostat – removal, testing and refitting

1   If the cooling system is hot, release the expansion tank cap before unscrewing the coolant filler cap on the cylinder head.
2   With the filler cap removed from the coolant filler neck on the

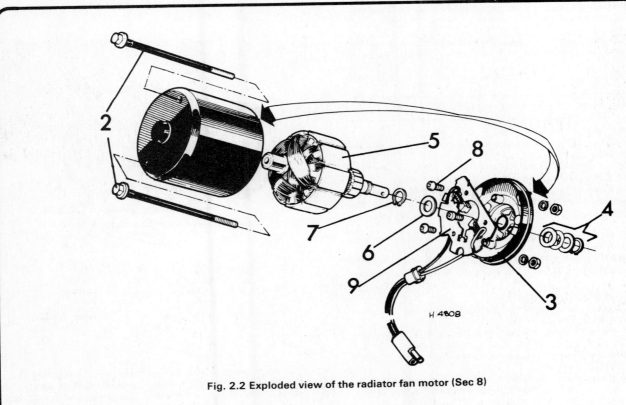

Fig. 2.2 Exploded view of the radiator fan motor (Sec 8)

2	Tie-bolts	6	Thrust washer
3	End cover	7	Circlip
4	Circlip and washers	8	Brush carrier screws
5	Armature	9	Brush carrier

9.2 Removing thermostat

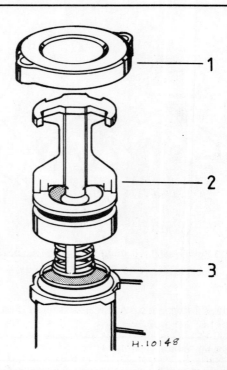

Fig. 2.3 Thermostat withdrawn from coolant filler neck (Sec 9)

1	Filler cap	3	O-ring seal
2	Thermostat		

cylinder head, pull the thermostat directly upwards and remove it (Fig. 2.3) (photo).

3   To test the thermostat, suspend it in a container of water and heat the water until the thermostat opens. By inserting a thermometer into the water, check the opening temperature of the thermostat and if this varies considerably from the figure stamped on it, fit a new one. Always renew the thermostat if it is stuck open or if one of incorrect rating (see Specifications) has been fitted by a previous owner. The fitting of a thermostat of incorrect rating will cause either overheating or cool running of the engine, slow warm up and an inefficient car interior heater.

4   Before fitting the thermostat, wipe the inside of the coolant filler neck clean and fit a new O-ring seal if the original one has deteriorated.

10 Drivebelts – tensioning, removal and refitting

Alternator/coolant pump drivebelt

1   The drive is taken from a twin grooved pulley on the front end of the crankshaft.

2   To tension the belt, loosen the alternator upper pivot bolt and the two bolts on the adjuster link at the base of the alternator.

3   Move the alternator away from the engine until the drivebelt is taut. When carrying out this operation have the bolts just tight enough to allow the alternator to pivot stiffly. Apply leverage only at the drive end bracket.

4   Nip up the bolts and then apply moderate thumb pressure to the centre of the longest run of the belt between two pulleys.

5   The belt should deflect 0.25 in (6.0 mm). Adjust the position of the alternator as necessary to achieve this, then tighten all bolts fully.

Power steering pump drivebelt

6   The drive is taken from one of two pulleys fitted to the front end of the coolant pump.

7   To tension the belt, slacken the union on the pump high pressure pipe.

8   Slacken the pump pivot and support bracket bolts and the adjuster link screws.

9   Using hand pressure pull the pump to tension the drivebelt. Nip up the bolts and apply moderate thumb pressure to the centre of the longest run of the belt between two pulleys.

10   The belt should deflect 0.25 in (6.0 mm). Adjust the position of the pump as necessary to achieve this, then tighten all bolts fully.

11   To remove a drivebelt, the alternator or power steering pump must be moved to the full extent of its travel to fully slacken the drivebelt. Slip the drivebelt over the pulley rim. If this proves difficult, turn the

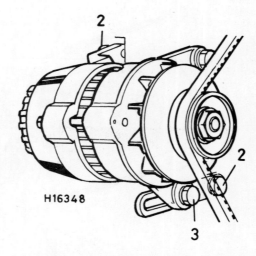

Fig. 2.4 Alternator mounting bolts (Sec 10)

2	Pivot mounting bolts	3	Adjuster link bolt

crankshaft pulley while prising the belt over one of the pulley rims.

12   If the alternator drivebelt is to be removed, then as all cars are equipped with power-assisted steering, the power steering drivebelt will have to be removed first (photo).

13   A drivebelt should be renewed if it shows signs of splitting, cracking or general deterioration.

14   If a new drivebelt is fitted, always check its tension after five minutes running and re-adjust if necessary as it may have stretched slightly.

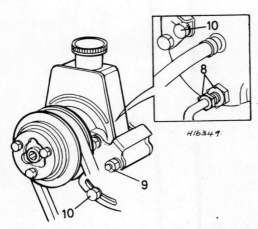

Fig. 2.5 Power-assisted steering pump drivebelt (Sec 10)

8   Pipe union nut
9   Pivot mounting bolt
10 Adjuster link screw

10.12 Drivebelts

11 Coolant temperature switch

1   The coolant temperature transmitter is screwed into the cylinder head just below the coolant filler/thermostat housing.
2   In the event of a faulty reading on the gauge, first check the electrical wiring connections.
3   If other instruments are also giving incorrect readings, the voltage stabilizer may be at fault (see Chapter 10) on the instrument panel.
4   Checking the temperature transmitter and gauge can only be satisfactorily carried out using special test equipment or by the substitution of new components. If the coolant temperature gauge is to be removed, refer to Chapter 10.
5   Consistently high or low readings on the coolant temperature gauge may indicate a faulty thermostat or one of incorrect temperature range.

12 Coolant filler/thermostat housing – removal and refitting

1   If the engine is hot, release the expansion tank cap before unscrewing the cap on the cylinder head coolant filler neck.

2   Refer to Chapter 1 and remove the timing belt cover.
3   Unscrew and remove the cap from the coolant filler housing.
4   Partially drain the coolant from the cylinder block and refit the drain plug.
5   Disconnect the coolant hose from the filler neck.
6   Unscrew and remove the screw which holds the coolant filler housing to the cylinder head and pull the housing straight out of its recess in the cylinder head.
7   Withdraw the thermostat and its O-ring seal.
8   Extract the filler housing O-ring seal.
9   Refitting is a reversal of removal. Do not overtighten the housing securing screw.

13 Coolant pump – removal and refitting

1   Drain the cooling system as described in Section 2.
2   Refer to Chapter 1 and remove the timing belt cover and the alternator drivebelt.
3   Extract the setscrews and pull off the coolant pump pulley.
4   Disconnect the coolant hose from the coolant pump.
5   Remove the timing belt tensioner pulley, noting that the securing

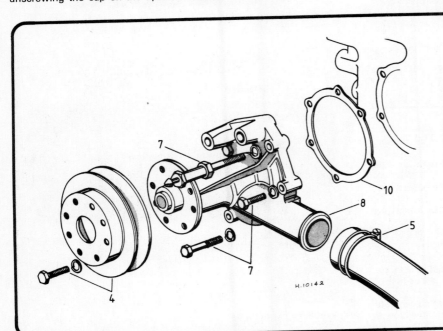

Fig. 2.6 Coolant pump (Sec 13)

4   Pulley
5   Coolant hose
7   Pump fixing bolt with
    timing pointer
8   Pump body
10 Gasket

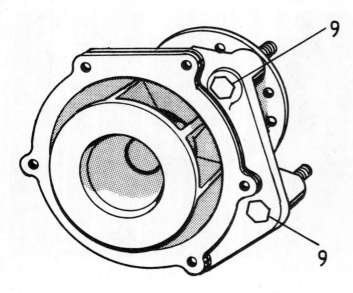

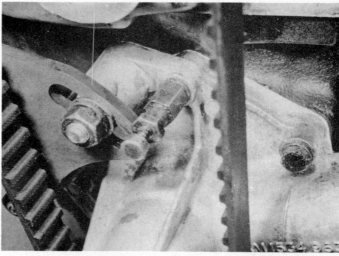

13.7 Timing belt tensioner nut and cover spigot at coolant pump

Fig. 2.7 Rear view of coolant pump (Sec 13)

9   Timing belt tensioner bolts

13.9 Coolant pump pulley

nuts are screwed onto floating bolts located in the rear of the pump.
6   Unscrew and remove the five bolts and the stud which hold the water pump to the block.
7   Ease the pump from the cylinder block and extract it from between the timing belt (photo).
8   Make sure that the two floating bolts on which the timing belt tensioner is mounted are removed from the rear of the coolant pump body (Fig. 2.7).
9   Peel away the coolant pump joint gasket. Unbolt and remove the pulley (photo).
10 A faulty or leaking pump should be renewed as a complete assembly, not repaired.
11 Refit the pump using a new joint gasket and making sure that the mating surfaces are clean.
12 Use thick grease to hold the two timing belt tensioner pulley bolts in position in the coolant pump body.
13 Tension the timing belt as described in Chapter 1, Section 7.
14 Tighten all bolts to the specified torque.
15 Tension the drivebelt for the alternator and power steering pump (if fitted) as described in Section 10.
16 Refill the cooling system (Section 4).

14 Heating and ventilation system – description

1   Heat for the system is provided by the engine coolant passing through the heater matrix.
2   Air can be delivered to the car interior, windscreen or side windows according to the setting of the controls.
3   Fresh air is drawn in through the grille which is mounted just forward of the windscreen.
4   Stale air is extracted through flap valves located in the lower part of the tailgate inner panel (photo).
5   Foot level ventilation is provided by outlets at the ends of the parcels shelf with control being provided by a slide control lever below the facia fresh air outlets.
6   Air from the heating and ventilation system is boosted by a three-speed blower fan which operates in conjunction with the position of the air control lever.
7   The heater control panel incorporates four levers which regulate the air flow and distribution, the fresh air intake volume and the blower speed.

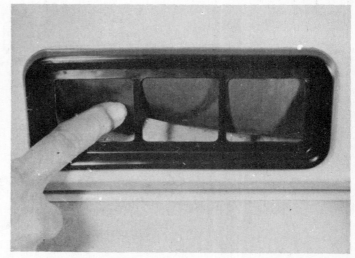

14.4 Stale air extractor flap

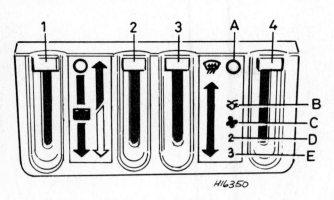

H16350

Fig. 2.8 Heater controls (Sec 14)

1   Face level air flow control   A   Heater off
    (volume)   B   Fresh air intake open
2   Temperature control lever   C   Blower slow speed
3   Air distribution (directional)   D   Blower normal speed
4   Fresh air volume/blower   E   Blower fast speed
    speed

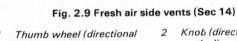

H16351

Fig. 2.9 Fresh air side vents (Sec 14)

1   Thumb wheel (directional   2   Knob (directional flow –
    flow – horizontal)   vertical)
                                    3   Foot level vent control

15 Heater – removal and refitting

1   Disconnect the battery.
2   Drain the cooling system (Section 2).
3   Remove the centre console (Chapter 12).
4   Remove the radio (Chapter 10).
5   Remove the heater cover. Do this by pulling off the heater control lever knobs, and extending the screws from below the centre vents and at the rear and base of the cover. Pull the cover forward and disconnect the cigar lighter and lamp leads (photos).
6   Remove the facia panel (Chapter 12).
7   Release the heater upper mounting bolt (photo).

8   Move the earth lead aside and detach the upper mounting bracket from the heater.
9   Detach the wiring harness from the top of the heater and move the harness aside.
10 Separate the radio feed wire from the wiring harness and remove the strap from the right-hand demister duct and then pull out the aerial and radio harness (photo).
11 Remove the face level demister ducts and disconnect the wiring harness multi-pin plug on the top of the heater (photo).

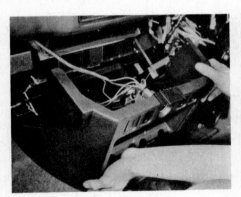

15.5A Removing heater cover

15.5B Heater cover removed

15.7 Heater upper mounting bolt

15.10 Heater control linkage

15.11 Face level demister duct

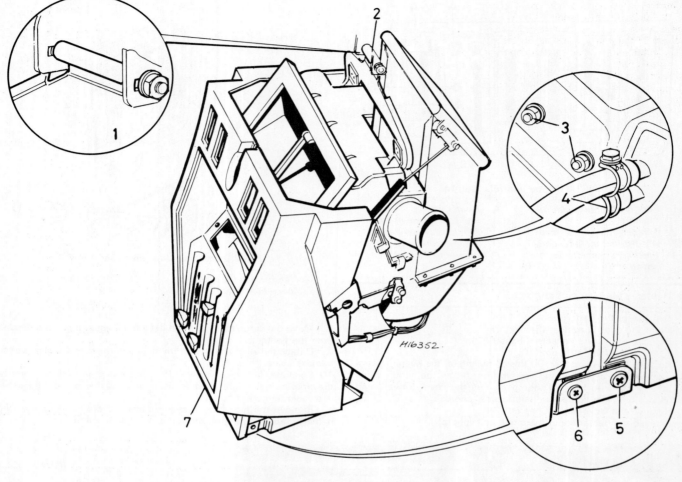

Fig. 2.10 Heater unit (Sec 15)

1	*Upper mounting bracket*	3	*Bulkhead fixing nuts*	6	*Console mounting bracket*
2	*Upper mounting bolt*	4	*Hose clips*		*screws*
		5	*Floor mounting screws*	7	*Heater cover*

12 Release the adhesive tape and detach the wiring harness from the heater and move it aside.

13 Disconnect the heater control lighting wire.

14 Extract the screws which hold the heater to the floor bracket.

15 Move the windscreen washer bottle aside and disconnect the heater hoses from the heater pipe stubs. Plug the pipe outlets to prevent spillage of coolant on the carpets.

16 Unscrew the heater fixing nuts on the engine compartment bulkhead and withdraw the heater from the car. Note the location of the drain tube.

17 Refitting is a reversal of removal.

18 Fill the cooling system.

19 Reconnect the battery.

20 Adjust the controls if necessary as described in Section 17.

16 Heater – dismantling and reassembly

1 With the water unit removed from the car, pull the electrical leads through the grommet on the air inlet box. Remove the securing screws and withdraw the air inlet box.

2 Prise off the casing flange spring clips and extract the self-tapping screws.

3 Separate the casings and lift out the blower motor. Pull the fan from the motor shaft.

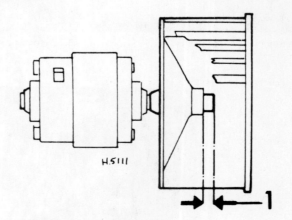

Fig. 2.11 Heater blower fan to motor shaft fitting diagram (Sec 16)

1 = 0.197 in (5.0 mm)

4   If the matrix is to be removed, lift off the face level flap, raise the air mixing flap and disconnect the operating link from the lever.
5   Withdraw the seal from the matrix casing and then, holding the pipe bracket and pipes push the pipes up through the heater casing and extract the matrix from the casing.
6   A fault in either the motor or matrix is best rectified by installation of a new component but if the matrix is blocked it is worthwhile reverse flushing it or using a reliable cleansing agent. If the matrix is leaking, do not waste your time trying to solder it but obtain a new or reconditioned unit.
7   Reassembly is a reversal of dismantling, but observe the following points:

    (a) Fit new pipe seals in the matrix
    (b) Make sure that the rubber mounting strips are stuck securely to the motor casing
    (c) Refit the fan so that $\frac{3}{16}$ in (5.0 mm) of the shaft projects through the fan hub (Fig. 2.11)
    (d) Locate the motor so that its tag engages with the lower half of the heater casing

8   Clip and screw the sections of casing together.

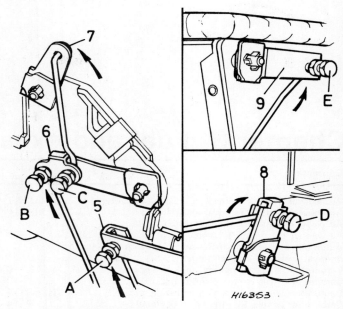

---

17 Heater controls – adjustment

1   The following operations should only be necessary if the heater controls are not operating correctly or after overhaul and dismantling.
2   Adjustment of the various controls does not have to be carried out in any particular sequence.
3   Remove the centre console (Chapter 12), the radio (Chapter 10).
4   Pull off the heater control lever knobs.
5   Remove the heater cover (Section 15).
6   Move the heater control levers upwards to the limit of their travel and then release the lockscrews at the flap trunnions. Rotate the valve flaps to their fully closed position and then tighten the lock screws.

Fig. 2.12 Heater controls and linkage (Sec 17)

5 Face level air flap lever	8 Air distribution flap lever
6 Air temperature intermediate lever	9 Fresh air flap lever
7 Air temperature flap lever	A, B, C, D, E Locking screws

---

18 Fault diagnosis – cooling system

Symptom	Reason(s)
Heat generated in engine not being successfully disposed of by radiator	Insufficient water in cooling system Drivebelt slipping (accompanied by a shrieking noise on rapid engine acceleration) Radiator core blocked or radiator grille restricted Bottom water hose collapsed, impeding flow Thermostat not opening properly Ignition advance and retard incorrectly set (accompanied by loss of power and perhaps misfiring) Carburettor incorrectly adjusted (mixture too weak) Exhaust system partially blocked Oil level in sump too low Blown cylinder head gasket (water/steam being forced down the expansion tank pipe under pressure) Engine not yet run-in Brakes binding
Too much heat being dispersed by radiator	Thermostat jammed open Incorrect grade of thermostat fitted allowing premature opening of valve Thermostat missing
Leaks in system	Loose clips on water hoses Top or bottom water hoses perished and leaking Radiator core leaking Thermostat gasket leaking Pressure cap spring worn or seal ineffective Blown cylinder head gasket (pressure in system forcing water/steam down expansion tank pipe) Cylinder wall or head cracked (internal leak)

Chapter 3 Fuel and exhaust systems

Contents

Specifications

System type .... Single or twin SU carburettor, rear mounted fuel tank with electric fuel pump. Temperature controlled type air cleaner

Fuel tank capacity .... 16.0 Imp gal (73.0 litres)

Fuel octane rating .... 97 Ron (4 star)

Carburettor damper
Oil type/specification .... Multigrade engine oil, viscosity SAE 10W/40, 15W/40 or 15W/50, to API SE/SF (Duckhams QXR, Hypergrade or 10W/40 Motor Oil)

Carburettor (1.7 engine and 2.0 HL engine up until 1982)
Type .... Single SU HIF 44
Identification number:
    Manual transmission .... FZX 1386
    Automatic transmission .... FZX 1387
Piston spring colour .... Red
Jet size .... 0.100 in
Needle .... BFA
Idle speed .... 650 rev/min
Fast idle speed .... 1100 rev/min
Exhaust gas CO content .... 1.5 to 3.5%

Carburettor (2.0 engine except HL to 1982)
Type .... Twin SU HIF 44 with ASU
Identification number:
    Manual transmission .... FZX 1355
    Automatic transmission .... FZX 1356
Piston spring colour .... Red
Jet size .... 0.100 in
Needle .... BFK
Idle speed .... 750 rev/min
Exhaust gas CO content .... 1.5 to 3.5%

Torque wrench settings	lbf ft	Nm
Carburettor to manifold nuts	19	25
Manifold nuts and bolts	18	24
Exhaust downpipe clamp nuts	16	22
Exhaust U-bolt nuts	18	24
Mounting bracket bolts	16	22

1 Description and maintenance

1   The fuel system consists of a rear-mounted fuel tank, an electrically-operated fuel pump, SU carburettor and a temperature controlled air cleaner (photos).
2   A single carburettor is fitted to all 1.7L and 1.7HL models and to 2.0HL models up until 1982.
3   Twin carburettors are fitted to 2.0HL models from 1983 and to all 2.0HLS and Vanden Plas versions.
4   Regular maintenance should include renewal of the air cleaner element, topping up the carburettor damper and checking the security of the fuel lines and hoses.

2 Air cleaner element – renewal

Single carburettor

1   At the intervals specified in Routine Maintenance, renew the air cleaner element. In particularly dusty conditions, the element should be changed more frequently.
2   Unscrew the air cleaner cover centre nut, remove the cover and take out the element and discard it (photos).

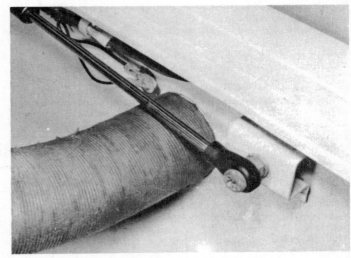

1.1A Cold air intake – temperature controlled air cleaner

1.1B Hot air intake – temperature controlled air cleaner

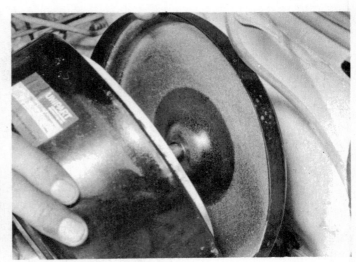

2.2A Removing air cleaner lid

2.2B Removing air cleaner element

2.2C Air cleaner sealing ring

3   Clean the interior of the air cleaner casing and cover. Check that the sealing ring and washer are in good condition and fitted correctly.
4   Refit the new element and the cover.

Twin carburettors

5   Disconnect the air intake duct from the air cleaner.
6   Release the breather and overflow hoses from their clip on the air cleaner casing.
7   Unscrew the two wing nuts from the top of the casing and withdraw the air cleaner from the carburettors.
8   Prise up the plastic tabs and remove the cover.
9   Remove the elements and discard them and thoroughly wipe out the interior of the casing.

10 Fit the new elements and check that the carburettor and wing nut sealing washers are in good condition and correctly located.
11 Refit the cover and the air cleaner to the carburettors.

3   Air cleaner temperature control valve

1   Occasionally, check the operation of the automatic temperature control valve. To do this, have the engine cold and note the position of the valve.
2   Depress the valve and then release it. The valve should return at once to its original position. If it does not, the valve assembly should be renewed.

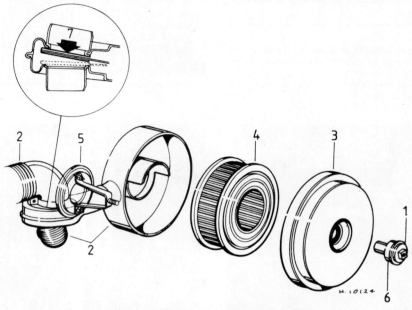

Fig. 3.1 Air cleaner (single carburettor) (Sec 2)

| 1 | Nut | 3 | Cover | 5 | Sealing washer |
| 2 | Air intake hose | 4 | Element | 6 | Sealing washer |

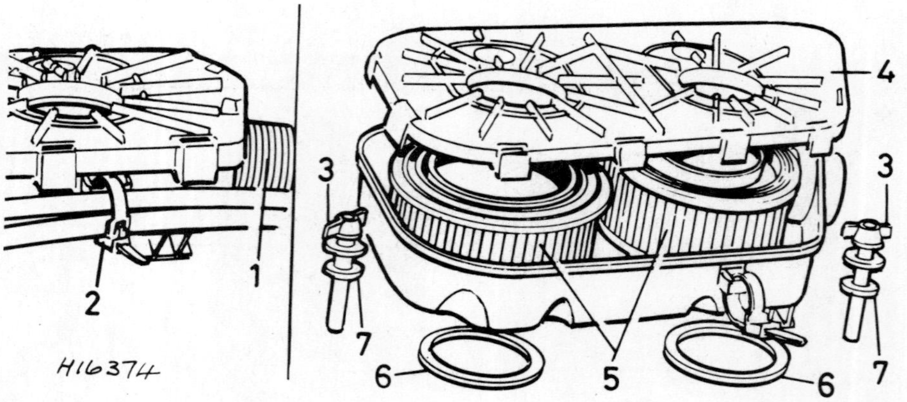

Fig. 3.2 Air cleaner (twin carburettor) (Sec 2)

1	Air intake hose	4	Cover	6	Sealing rings
2	Hose clip	5	Filter elements	7	Sealing rings
3	Wing nut				

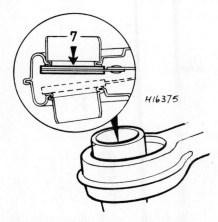

Fig. 3.3 Air cleaner valve (Sec 3)

7   Valve flap

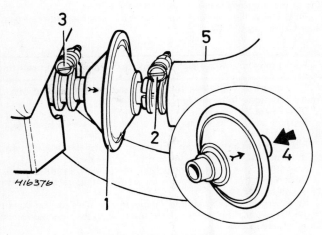

Fig. 3.4 ASU filter (Sec 4)

1	Filter	4	Direction of air flow (large arrow)
2	Inlet hose clamp		
3	Clamp	5	Hot air inlet hose

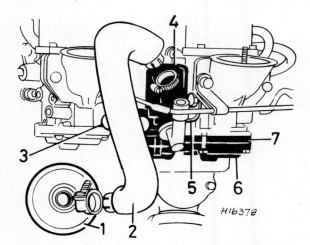

Fig. 3.5 ASU attached to left-hand carburettor (Sec 5)

1	Air intake filter	5	Support bracket
2	Hot air inlet hose	6	Carburettor float chamber
3	Outlet hose		bottom cover
4	ASU	7	Adaptor

4   Hot air intake filter (automatic starting unit-ASU) – description and cleaning

1   This device used only on twin carburettor versions (left-hand unit) reacts to carburettor air intake temperature, and to intake manifold depression in order to provide a richer mixture for cold starting.
2   When the air intake temperature reaches 104°F (40°C) the provision of enriched mixture ceases.
3   The ASU filter should be cleaned at the same time as the air cleaner is removed for element renewal.
4   With the air cleaner removed, slacken the hose clip and pull the hot air inlet hose from the filter.
5   Slacken the remaining clip and pull the filter from the adaptor on the manifold.
6   Wash the device thoroughly in petrol and blow it through with compressed air or air from a tyre pump against the directional flow arrowed on the unit.
7   Refit the filter so that the directional arrow points away from the adaptor on the manifold.

5   Automatic starting unit (ASU-twin carburettors) – testing, removal, overhaul and refitting

Testing
1   Should it be suspected that the ASU is not operating correctly, carry out the following test.
2   Remove the air cleaner and check that the ASU hoses are secure and that the fuel level in the left-hand carburettor is correct (see Section 14).
3   If the engine is cold, operate the starter motor for five seconds. If the engine starts immediately, switch off again after a five second period of idle.
4   Disconnect the outlet hose and check that the outlet port is moist with petrol. If not, refit the outlet hose, partially block the inlet hose and crank the engine for five seconds.
5   If the engine starts immediately, switch off after a five second period of idle. If the outlet port is now wet, the ASU is defective.
6   If the engine is hot, check that the inlet port is not open. If it is, the ASU is defective. If difficult hot starting is experienced, remove the outlet hose and blank off the manifold. If the engine now starts, the ASU is defective.

Removal
7   To remove the ASU, disconnect the battery, remove the air cleaner and intake assembly.
8   Remove the hot air inlet hose.
9   Disconnect the outlet hose from the manifold adaptor.
10   Extract the support bracket screw noting the washer under the bracket. Mop up released fuel with a rag.
11   Extract the screws which hold the bottom cover to the carburettor. Withdraw the ASU complete with adaptor plate.
12   Remove the outlet hose.

Overhaul
Note: *First ascertain whether spare parts are available. If not, the ASU must be renewed as a complete assembly if faulty.*
13   The limit of overhaul is renewal of the diaphragms.
14   Scribe a line across the edges of the body and the covers to ensure correct alignment at reassembly.
15   Extract the screws and separate the bottom covers. Note the locations of the hollow dowels.
16   Remove the diaphragm, spring and pushrod.
17   Remove the second diaphragm from the bottom cover.
18   Blow out all passages with air from a tyre pump.
19   Commence reassembly by checking that the dowel is in position and then locate the bottom cover diaphragm with its gasket side uppermost (visible) on the cover.
20   Fit the remaining hollow dowels, the pushrod and coil spring.
21   Fix the bottom cover to the body using the screws.

Refitting
22   Locate new seals on the adaptor plate and bottom cover and fit the ASU to the carburettor, but leave the screws finger tight.
23   Fit the plain washer and the support bracket leaving its screw finger tight.

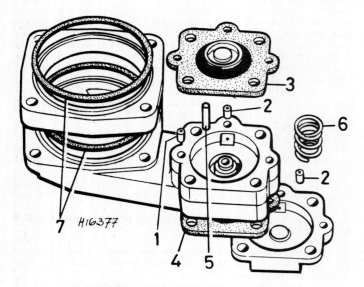

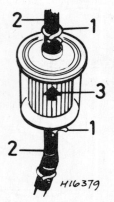

Fig. 3.7 In-line fuel filter (Sec 6)

1 Hose clip	*3 Directional flow arrow*
2 Fuel hose	

clips at the filter and pull the hoses from the filter.

3 Fit the new filter making sure that the arrow marked on it points towards the carburettor.

Fig. 3.6 Exploded view of ASU (Sec 5)

1 Hollow dowel – fuel	*5 Pushrod*
2 Hollow dowel – air	*6 Return spring*
3 Diaphragm	*7 Seals*
4 Diaphragm	

24 Tighten the bottom cover screws then the support bracket to ASU screws then the carburettor screws in that order.

25 Connect the outlet hose and inlet hose, refit the air cleaner and reconnect the battery.

7 Fuel pump and fuel level transmitter – removal and refitting

1 The electrically-operated fuel pump and transmitter unit may be removed together from the fuel tank.

2 Disconnect the battery.

3 Syphon any fuel in the tank into a metal can which has a secure cap.

4 Raise the rear of the car and support it on axle stands. Clean away any dirt from the pump/transmitter mounting plate.

5 Disconnect the electrical plug from the pump/transmitter socket.

6 Disconnect the fuel hose from the pump outlet, catching the fuel which will be released.

7 Release the pump/transmitter retaining ring. This can be done by using a flat piece of metal up against the tabs on the ring and rotating it in a clockwise direction.

6 Fuel filter – renewal

1 A disposable type fuel filter is used in the fuel feed line of twin carburettor models.

2 At the intervals specified in Routine Maintenance, release the hose

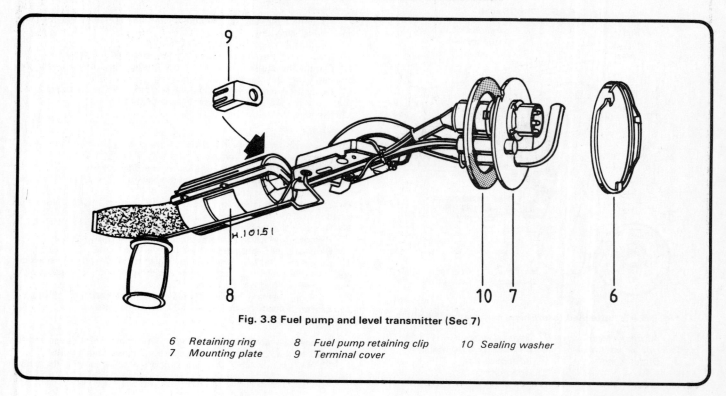

Fig. 3.8 Fuel pump and level transmitter (Sec 7)

6 Retaining ring	*8 Fuel pump retaining clip*	*10 Sealing washer*
7 Mounting plate	*9 Terminal cover*	

8   Carefully withdraw the pump/transmitter assembly out of the fuel tank.

9   Press the fuel pump from its circular retaining clip and then pull it from the fuel outlet hose. Disconnect the electrical leads from the pump.

10 The pump and transmitter can be renewed independently.

11 Refitting is a reversal of removal but make sure that the pump electrical leads are correctly connected: RED to positive terminal, BLACK to negative terminal. Push the pump into its securing clip so that the positive (+) connector is nearest the clip.

12 The corbin clip on the hose should have its open ends pointing inwards.

13 Fit a new rubber sealing ring under the pump/transmitter mounting plate if there is any doubt about the condition of the original one.

8   Fuel tank — removal, repair and refitting

1   Disconnect the lead from the battery negative terminal.

2   Remove the fuel filler cap, and syphon fuel into a suitable container.

3   Raise the rear of the car and support it under the rear jacking points.

4   Detach the lead from the tank transmitter unit and remove the lead also from the securing clips on the side of the fuel tank.

5   Disconnect the fuel outlet pipe from the tank transmitter unit and catch any fuel which drains in a suitable container.

6   Detach the fuel pipe from the two clips on the side of the tank.

7   Support the weight of the fuel tank on a jack and insulating wooden block and then remove the five screws and plates which secure the tank to the bodyframe (photo).

8   Lower the fuel tank, carefully extracting the filler pipe from the body.

9   The tank pump transmitter unit can be released by turning it with a suitable lever.

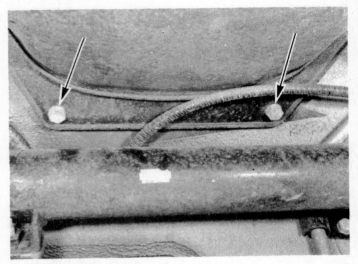

8.7 Fuel tank bolts

10 Never attempt to repair a leaking fuel tank by soldering or welding. Even when empty, explosive gases will remain unless the tank is steamed out or flushed through with boiling water for at least an hour. Have the tank professionally repaired or fit a new one.

11 If the tank contains a quantity of sludge or sediment, use several changes of paraffin, shaking the tank vigorously to dislodge it but remove the transmitter unit first to prevent damaging it.

12 When refitting the transmitter unit, use a new sealing gasket.

13 Refitting is a reversal of removal.

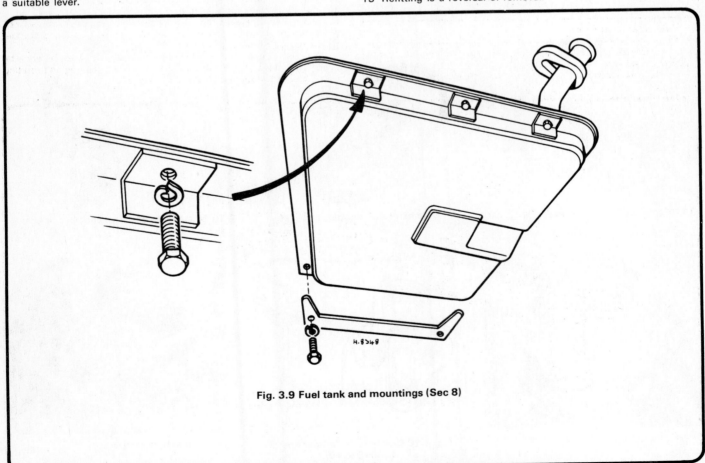

Fig. 3.9 Fuel tank and mountings (Sec 8)

9 Fuel contents gauge and transmitter unit

1   In the event of the fuel contents gauge reading incorrectly, first check the security and insulation of the wiring between the gauge and transmitter.

2   If the coolant temperature and fuel gauges are both giving incorrect readings, the instrument voltage stabiliser may be at fault (see Chapter 10).

3   Checking the transmitter and the gauge can only be satisfactorily carried out using special test equipment or by substitution of new units.

4   Removal of the tank transmitter unit is described in Section 7, while removal of the gauge is covered in Chapter 10.

10 Carburettor – description and maintenance

1   The carburettors are of horizontal draught type (photos).

2   Where a single carburettor is fitted, a manual choke (cold start) is used. On twin carburettor versions, an automatic choke is used.

3   Maintenance consists of keeping the units clean externally with the control linkage pivots lubricated and the mounting nuts tight.

4   At the intervals specified in Routine Maintenance, unscrew the cap from the top of the carburettor suction chamber and withdraw the cap with damper.

5   Pour in clean engine oil until it reaches the top of the hollow piston rod.

6   Refit the damper and cap. To do this will require gentle downward pressure on the cap against the resistance of the oil (photo).

7   At the specified intervals, renew the float chamber air filter which is a push fit.

11 Carburettor (single) – running adjustments

1   As is the case with most modern vehicles, the carburettor is set in production and is regarded as non-adjustable.

2   However, it is rare to find a vehicle which does not require slight adjustment of the idle speed mixture or fast idle once the engine has run-in. Further adjustment may also be required as the engine characteristics change (carbon build up or internal wear) or after the carburettor has been overhauled.

3   The idle speed and mixture screws are sealed during production and their tamperproof plugs must be prised out or drilled carefully and extracted using a self-tapping screw before the following operations can be carried out.

4   Where regulations require it, new red tamperproof plugs must be fitted on completion.

5   A tachometer must be connected to the engine and for really accurate mixture adjustment, an exhaust gas analyser (CO meter) should be used.

Idle speed and mixture adjustment

6   Have the engine at normal operating temperature with the engine switched off.

7   Remove the air cleaner and the carburettor piston damper.

8   With the finger, lift the piston and check that it drops freely onto its bridge. If it does not it must be removed and cleaned with solvent.

9   Check that the throttle cable has a small amount of free play and that there is a gap between the end of the fast idle screw and the cam.

10 Connect the tachometer, start the engine and allow it to idle in neutral (manual) or P (automatic).

11 Increase the engine speed to 2500 rev/min for a period of thirty seconds and then return it to idle.

12 Immediately turn the idle speed screw until the engine idles at the specified level.

10.1A Carburettor – throttle lever side

10.1B Carburettor – throttle valve plate side

10.1C Carburettor – float chamber air filter side

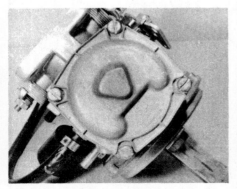

10.1D Carburettor – float chamber side

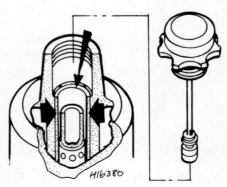

Fig. 3.10 Carburettor damper and oil level (Sec 10)

10.6 Carburettor damper

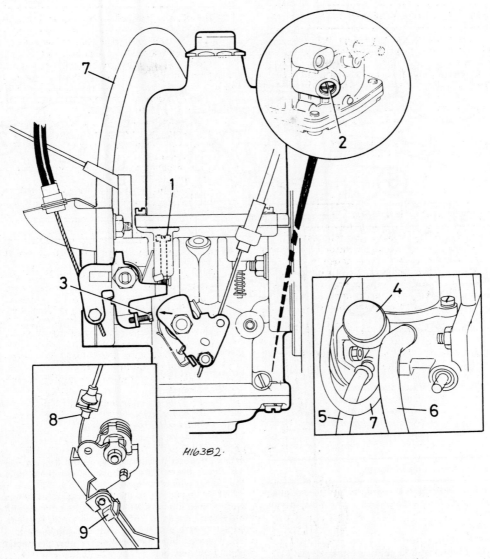

Fig. 3.11 Carburettor adjustment points and connections (Sec 11)

1	Idle speed screw	3	Fast idle screw (single	5	Fuel feed hose	8	Accelerator cable adjuster
2	Idle mixture screw		carburettor)	6	Engine breather hose	9	Kickdown cable (automatic
		4	Float chamber air filter	7	Additional weakening device		transmission)

13 Now turn the mixture screw until the fastest idling speed is obtained. Now turn the screw in an anti-clockwise direction (weaken) until engine speed just starts to drop. Turn the screw clockwise (enrich) until the highest idle speed is just attained.

14 Readjust the idle speed screw if necessary to regain the specified engine idle speed.

15 If the exhaust gas analyser is now connected, the CO level should be between 1.5 and 3.5%.

16 If any of the foregoing adjustments cannot be completed within three minutes of the clearing operation (increasing speed to 2500 rev/min for 30 seconds) then clear the engine again before proceeding with the adjustment.

Fast idle adjustment

17 Connect a tachometer to the engine which should be cold, preferably having stood overnight.

18 Pull the choke control out until the arrow on the cam is aligned with the centre of the fast idle screw.

19 Start the engine and check the speed. If it is not as specified (1100 rev/min), turn the fast idle screw as necessary.

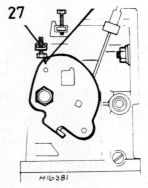

Fig. 3.12 Fast idle cam and screw (Sec 11)

26 Cam mark
27 Fast idle screw

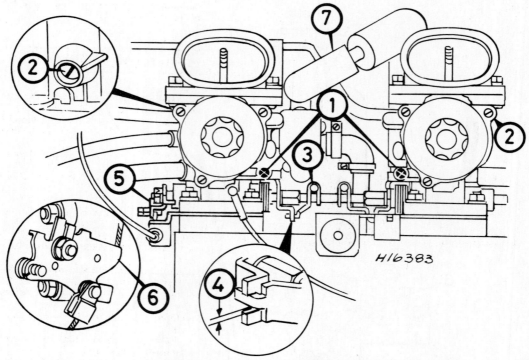

Fig. 3.13 Twin carburettor arrangement (Sec 12)

1	Idle speed screw	5	Throttle lever
2	Idle mixture screw	6	Progressive cam and
3	Interconnecting rod clamp		kickdown cable
4	Clearance (throttle lever		(automatic transmission)
	to fork)	7	ASU intake port

12 Carburettors (twin) – running adjustments

1   The information contained in Section 11 paragraphs 1 to 5 applies to the twin carburettor unit, but additionally an airflow meter (balancer) will be required.

Idle speed and mixture adjustment

2   Remove the air cleaners with hot and cold air intake ducts.
3   Remove both suction chambers amd check that each needle guide is flush with the piston face.
4   Turn each mixture adjusting screw to bring the jets flush with their bridges. Then turn each screw two turns clockwise so that both jets are lowered by the same amount.
5   Refit the suction chambers, but without their dampers.
6   Raise the pistons with the fingers and check that they fall freely onto their bridges.
7   Fill the hollow piston rods with engine oil and refit the dampers.
8   Slacken the throttle interconnection clamp and start the engine.
9   Run it at a fast idle speed until normal engine operating temperature is reached.
10   Turn the idle speed screws equally until the engine is running at 800 rev/min.
11   Connect an airflow meter and turn the idle speed screws as necessary to balance both carburettors (air intake flow equal).
12   Remove the airflow meter and turn each idle speed screw an exactly equal amount until the idle speed is as specified.
13   Now turn each mixture adjusting screw an equal amount (clockwise to enrich, anti-clockwise to weaken) until the engine idle speed is at its highest. Turn each screw anti-clockwise until the speed just starts to drop, then very slowly clockwise until the maximum idle speed is just attained.
14   Re-adjust the idle speed screws equally to achieve the specified idle speed.
15   Tighten the interconnection clamp making sure that the connecting rod has an endfloat of $\frac{1}{32}$ in (0.8 mm) and a slight clearance

exists between the throttle lever and fork on the rear carburettor.
16   Check the carburettor balance with the airflow meter and re-adjust the clamp where necessary.
17   It is unlikely that the foregoing adjustments can be completed within three minutes. After every three minutes have elapsed, increase the engine speed to 2500 rev/min for 30 seconds in order to clear the engine before proceeding with the adjustment.
18   If the exhaust gas analyser is now connected, then the CO level should be between 1.5 and 3.5%.
19   Switch off the engine and refit the air cleaner.

13 Carburettor – removal and refitting

Single carburettor

1   Remove the air cleaner and intake ducts.
2   Disconnect the fuel hoses from the carburettor and plug them.
3   Disconnect the crankcase ventilation and vacuum hoses.
4   Disconnect the throttle return spring and the accelerator cable, and the choke control cable (photo).
5   On cars equipped with automatic transmission, disconnect the kickdown cable.
6   Unscrew and remove the mounting nuts and lift the carburettor from the intake manifold. Note the gasket and shield (photo).
7   Before refitting the carburettor, clean away all remnants of old flange gasket. Fit a new gasket.
8   Locate the carburettor on the manifold and tighten the nuts to the specified torque.
9   Connect the control cables making sure that the accelerator and choke cables have some free movement. Adjust the kickdown cable as described in Chapter 6.
10   Reconnect all the hoses.
11   Top up the damper with engine oil, refit the air cleaner.
12   If the carburettor was dismantled, carry out the adjustments described in Section 11.

13.4 Carburettor connections

1 Choke control cable	4 Throttle control cable
2 Idle speed screw	5 Fast idle screw
3 Distributor vacuum pipe	

13.6 Carburettor flange gasket and shield

Twin carburettors

13 Remove the air cleaner and intake ducts.

14 Disconnect the fuel hoses from the carburettor and plug their ends.

15 Disconnect the accelerator cable.

16 On cars equipped with automatic transmission, disconnect the downshift cable.

17 From the ASU unit (Sections 4 and 5) release the clip and remove the hot sensor pipe.

18 Release the clip which holds the outlet hose to the intake adaptor.

19 Unscrew the carburettor mounting nuts and lift both units from the manifold. It is possible to remove each carburettor independently provided the interconnecting rod is first disengaged and in the case of the left-hand unit, the ASU unit withdrawn first.

20 Use new gaskets on clean flanges before refitting and ensure that a little free movement, $\frac{1}{8}$ in (4.0 mm), is provided when the accelerator cable is being connected.

21 Connect the kickdown cable (automatic transmission) and adjust as described in Chapter 6.

22 Top up the dampers with engine oil.

23 If the carburettors were dismantled, carry out the adjustments described in Section 12.

14 Carburettor (Single SU HIF 44) – overhaul

Dismantling

1   With the carburettor removed, clean away external dirt.

2   Unscrew the damper and pour out the oil.

3   Extract the screws and remove the suction chamber.

4   Remove the piston spring and carefully lift out the piston/jet needle assembly.

5   Release the small needle retaining grub screw and withdraw the needle, guide and bias spring.

6   Mark the relative position of the float chamber cover to the carburettor body and remove the cover. Mop up the fuel which will be released.

7   Remove the jet adjusting screw and spring and the jet adjusting lever retaining screw.

8   Withdraw the jet and lever.

9   Unscrew the float pivot spindle and remove the float, needle valve and valve seat.

10 Unscrew the jet bearing nut and withdraw the jet bearing.

11 Note the location of the fast idle return spring.

12 Remove the cam lever retaining nut and tab washer. Remove the cam and spring.

13 Remove the dust cap and the starter assembly. Withdraw the seal from the spindle.

14 Unscrew the throttle spindle nut and take off the tab washer, throttle lever and return spring.

15 It is not recommended that the throttle valve plate is removed from its spindle but if it must be dismantled, first note how the plate is located before extracting the retaining screws.

16 Remove the throttle spindle seals from the carburettor body.

17 Wash all components in solvent and examine for wear.

18 Obtain a repair kit which will retain all the renewal items including the seals.

Reassembly

19 Commence reassembly by inserting the throttle spindle and locating the throttle valve plate so that its chamfer is towards the piston. The overrun valve (when fitted) will be at the bottom. Leave the screws finger tight.

20 Actuate the throttle several times to align the valve plate and then tighten the plate screws. Lightly peen the ends of the screws, taking great care not to distort the plate or spindle.

21 Press in the spindle seals so that they are just below the surface of the carburettor body (photo).

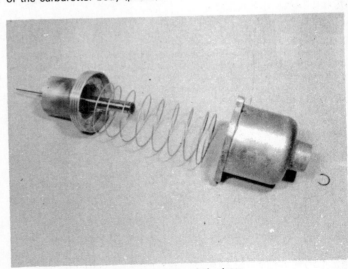

14.21 Suction chamber, coil spring and air piston

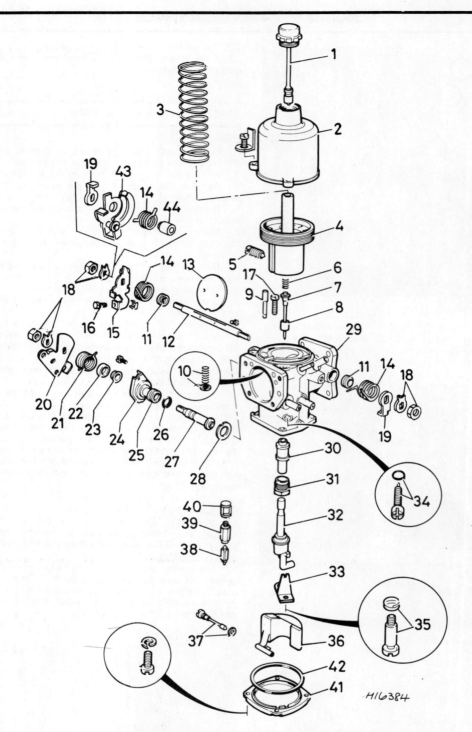

Fig. 3.14 Exploded view of SU HIF 44 carburettor (Sec 14)

1	Piston damper	13	Valve plate (butterfly)	24	Retaining plate	35	Jet retaining screw and spring
2	Suction chamber	14	Return spring	25	Cold start body		
3	Piston spring	15	Throttle lever	26	O-ring seal	36	Float
4	Piston	16	Idle speed screw	27	Cold start spindle	37	Float pivot and seal
5	Grub screw	17	Fast idle screw	28	Gasket	38	Fuel inlet needle valve
6	Needle bias screw	18	Spindle nut and tab washer	29	Carburettor body	39	Needle valve seat
7	Jet needle			30	Jet bearing	40	Fuel filter
8	Needle guide	19	Throttle lever return spring retainer	31	Jet bearing nut	41	Float chamber cover
9	Piston lifting pin			32	Jet assembly	42	Cover seal
10	Spring and circlip for lifting pin	20	Fast idle cam	33	Bi-metal jet lever	43	Progressive throttle lever (automatic transmission)
		21	Fast idle return spring	34	Jet adjusting screw and seal		
11	Throttle spindle seals	22	Dust cap			44	Spacer
12	Throttle spindle	23	Spindle seat				

14.22 Fitting suction chamber

14.26 Float

22 Fit the throttle spring and lever, screw on the nut and bend over the tab washer to lock the nut (photo).

23 Refit the starter assembly, spindle, seals and dust cap.

24 Refit the fast idle cam and associated components.

25 Screw in the jet bearing nut.

26 Reassemble the float and fuel inlet needle valve. Invert the carburettor and allow the float to rest under its own weight. Place a straight-edge across the centre of the float chamber and measure the gap (A) between the bottom of the straight edge and the surface of the float. This should be between 0.02 and 0.04 in (0.5 and 1.0 mm). If it requires correction, bend the float arm tab (photo).

27 Fit the jet and lever and associated parts. Set the jet flush with the carburettor bridge.

28 Fit the float chamber cover aligning the marks made at dismantling.

29 When fitting the needle and its guide to the piston, make sure that the engraved mark is towards the transfer holes as shown. Tighten the grub screw when the needle guide is flush with the face of the piston.

30 Refit the piston/jet needle assembly, the piston coil spring.

31 When refitting the suction chamber, do not twist it, but place it directly into position otherwise the spring will be 'wound up'.

32 Fill the damper tube with oil and screw in the damper.

33 Once the carburettor has been refitted, carry out the tuning operations described in Section 11.

15 Carburettor (twin SU HIF 44 with ASU) – overhaul

Dismantling

1 The operations are very similar to those described for the single carburettor in the preceding Section but observe the following differences:

2 If the left-hand unit is being overhauled, mark the position of the ASU adaptor on the float chamber cover before removing it.

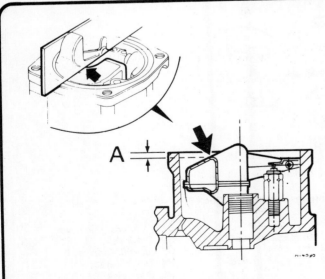

Fig. 3.15 Float setting diagram (Sec 14)

A = 0.02 to 0.04 in (0.5 to 1.0 mm)

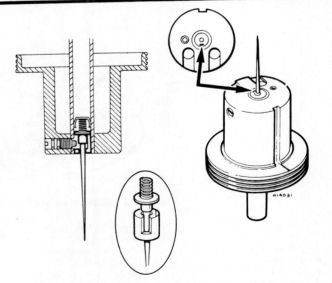

Fig. 3.16 Jet needle setting (Sec 14)

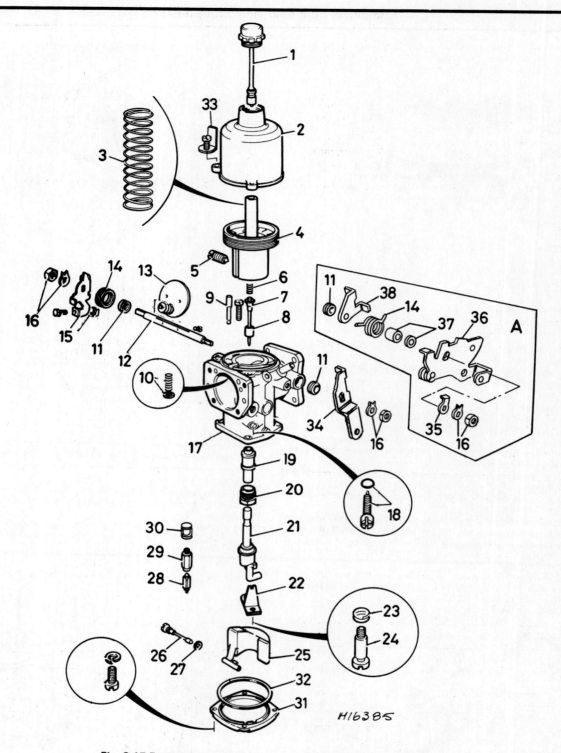

H16385

Fig. 3.17 Exploded view of carburettor (twin installation) (Sec 15)

1 Piston damper	11 Throttle spindle seals	21 Jet assembly	31 Float chamber cover
2 Suction chamber	12 Throttle spindle	22 Bi-metal jet lever	32 Cover seal
3 Piston spring	13 Valve plate (butterfly)	23 Jet spring	33 Identification tag
4 Piston	14 Throttle spring	24 Jet retaining screw	34 Throttle lever
5 Grub screw	15 Idle and interconnection	25 Float	35 Cam lever
6 Needle bias spring	lever	26 Float pivot	36 Progressive throttle lever
7 Jet needle	16 Tab washer and nut	27 Pivot seal	37 Spacer and washer
8 Needle guide	17 Carburettor body	28 Fuel inlet needle valve	38 Spring retainer
9 Piston lifting pin	18 Mixture screw and seal	29 Valve seat	A Applicable to automatic
10 Spring and circlip for	19 Jet bearing	30 Fuel filter	transmission models
lifting pin	20 Jet bearing nut		

3   The left-hand carburettor is fitted with an inter-connecting lever.
4   If the car is equipped with automatic transmission, a progressive cam assembly is fitted to the throttle spindle.

Reassembly

5   The operations are as described in the preceding Section, but note that the float level setting is 0.20 to 0.24 in (5.1 to 6.1 mm). Also, if it was the left-hand carburettor (with ASU) that was being overhauled then the float chamber cover must be refitted in the following way. Locate the sealing rings and the float chamber cover (mating marks in alignment) with the ASU adaptor, but leave the screws finger tight.
6   Locate the plain washer and fit the support bracket screw finger tight. Slacken the screw which secures the bracket to the ASU. Tighten all the screws in the following sequence:

1   Cover screws
2   Support bracket to ASU screw
3   Support bracket to carburettor screw

7   Once the carburettor has been refitted, carry out the tuning operations described in Section 12.

16 Accelerator pedal – removal and refitting

1   Remove the spring clip and disconnect the accelerator cable from the pedal under the facia panel.
2   Unscrew the two nuts which hold the accelerator pedal bracket to the brake servo studs.
3   Peel down the carpet behind the brake pedal and unbolt the accelerator pedal bracket from the brake pedal bracket.
4   Remove the accelerator bracket.
5   Refitting is a reversal of removal.

17 Accelerator cable – renewal

1   At the carburettor, disconnect the throttle return spring
2   Release the pinch-bolt and disconnect the inner cable from the lever trunnion at the carburettor.
3   Press the outer cable retainer from its slots in the support bracket.
4   Withdraw the cable assembly from the engine compartment rear bulkhead into the car interior.
5   Remove the spring clip and slip the cable end fitting out of the slot in the accelerator pedal.
6   Refitting is a reversal of removal, but make sure that the cable has a slight free movement, 0.079 in (2.0 mm), before the accelerator pedal moves the throttle lever on the carburettor.

18 Choke control (single carburettor)

1   Disconnect the cable from the carburettor.
2   If only the inner cable is to be removed, withdraw it by pulling the control knob.
3   If the complete inner and outer cable assembly must be removed, withdraw the left-hand shroud from the upper steering column.
4   Unscrew the control knob bezel locknut.
5   On models with a choke ON warning lamp switch, disconnect the wiring plug, remove the switch clip and release the clamp screw. Remove the switch.
6   Pull the complete cable assembly into the car interior.
7   Refitting is a reversal of removal, but make sure that the cable has a slight free movement.
8   Where a warning switch is fitted, make sure that the switch plunger and pips locate correctly and that the orange/green lead is connected nearest the control knob.

19 Manifolds – removal and refitting

1   Remove the carburettor from the intake manifold and move it to one side.
2   Disconnect the brake servo vacuum pipe from the intake manifold.
3   Disconnect the distributor vacuum advance pipe from the intake

Fig. 3.18 Accelerator cable to pedal connection (Sec 16)

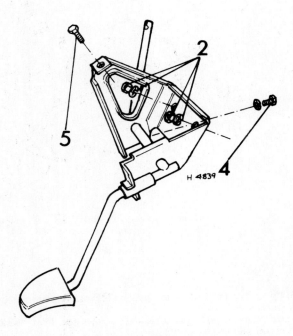

Fig. 3.19 Accelerator pedal assembly (Sec 16)

2   Servo mounting nuts   5   Setscrew
4   Setscrew

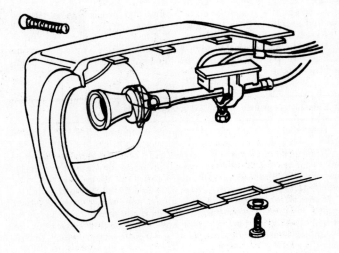

Fig. 3.20 Choke control and warning switch (Sec 18)

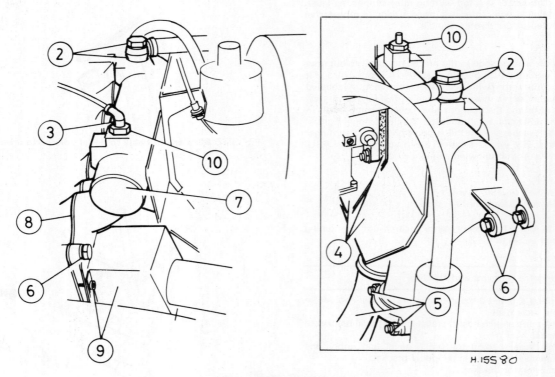

Fig. 3.21 Manifolds (Sec 19)

2   Vacuum servo connection   5   Exhaust downpipe clamps   8   Flange gasket
3   Distributor vacuum pipe   6   Manifold bolts   9   Hot box assembly
4   Heat shield   7   Manifold   10 Union adaptor

manifold. On left-hand drive cars, the location of these vacuum take off points is reversed.

4   On cars with manual transmission, remove the heat shield, support bracket and insulation block.

5   On cars with automatic transmission, place the heat shield to one side to avoid having to disconnect the attachments.

6   Remove the exhaust pipe to manifold clamps.

7   Before the manifold bolts can be unscrewed, the air cleaner hot air box must be removed. The front of the box is held to its backplate by two self-tapping screws which must be extracted.

8   Unscrew the manifold bolts evenly and progressively. Lift the manifold assembly from the cylinder head.

9   Remove the manifold gasket.

10 Refitting is a reversal of removal but always use a new gasket and sealing rings on the exhaust pipe flanges. Tighten the bolts to the specified torque.

20 Exhaust system – removal and refitting

1   The exhaust system fitted as original equipment is of one-piece construction and incorporates a spring-loaded balljoint.

2   To remove the complete system is a problem due to the fact that there must be a drop of four or five feet under the rear of the car in order to pull the pipe down sufficiently far to enable the twin downpipes to pass between the engine and the body member.

3   If the exhaust pipe is being removed for renewal then the best thing to do is to hacksaw through the pipe and remove it in sections. The ball coupling is of swaged type and cannot be dismantled. Pattern exhaust systems are obtainable in separate sections which makes refitting easier.

4   Whichever removal method is being used carry out the following operations.

5   Disconnect the downpipes from the manifold.

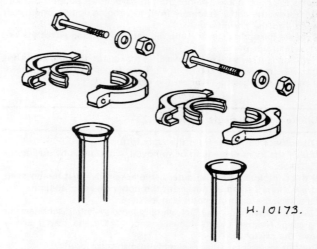

Fig. 3.22 Exhaust downpipe clamps and seals (Sec 19)

6   Remove one bolt from the end of the reinforcement strut which runs between the inboard ends of the lower track control arms. Release the bolt at the opposite end and swivel the strut downward (photos).

7   Disconnect the exhaust system flexible mountings (photos).

8   Reassemble the new components but do not tighten any pipe clamps until the system has been fitted on its mountings and the silencer and expansion box checked for correct alignment.

9   Use new asbestos seals or compound at the manifold to downpipe clamps.

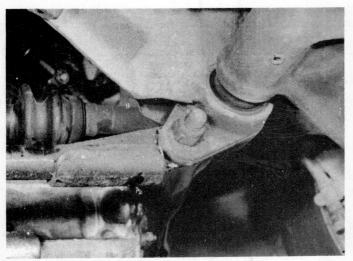

20.6 Front suspension reinforcement strut

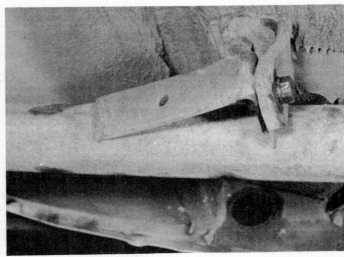

20.7A Exhaust rear mounting

20.7B Exhaust mounting at silencer

20.7C Exhaust centre mounting

Fault diagnosis appears overleaf

21 Fault diagnosis – fuel system and carburation

Symptom	Reason(s)
Fuel consumption excessive	Air cleaner choked giving rich mixture Leak from tank, pump or fuel lines Float chamber flooding due to incorrect level or worn needle valve Carburettor incorrectly adjusted Idling speed too high Incorrect valve clearances
Difficult starting	Lack of fuel Incorrect fuel level in float chamber Incorrectly adjusted mixture screw Air leak on induction side Faulty operation of temperature controlled air cleaner Overrun valve stuck (twin carburettors) Faulty ASU (twin carburettors)
Flat spot or lack of power	Low octane fuel Clogged fuel filter (twin carburettors) Incorrect fuel level in float chamber Damper requires topping up Incorrectly adjusted mixture screw Air leak on induction side Overrun valve stuck (twin carburettors) Faulty ASU (twin carburettors) Faulty fuel pump
Poor or erratic idling	Weak mixture Air leak on induction side Leak in manifold vacuum hoses Faulty throttle disc overrun valve (LH carburettor on twin units)

Chapter 4 Ignition system

Contents

Specifications

System type
Battery, coil and mechanical breaker distributor driven from overhead camshaft

Firing order
1–3–4–2

Ignition timing
17H engine	16° BTDC at 1500 rev/min*
20H engine:	
Single carburettor	12° BTDC at 1500 rev/min*
Twin carburettor	18° BTDC at 2000 rev/min*

Vacuum advance pipe disconnected

Distributor
Type	Lucas 48D4
Rotor rotation	Anti-clockwise
Contact breaker gap	0.014 to 0.016 in (0.35 to 0.40 mm)
Dwell angle	49 to 59°
Condenser capacity	0.18 to 0.25 microfarad

Ignition coil
Type	AC Delco 9977232 or Ducellier 520035A

Spark plugs
Type	Unipart GSP 361
Electrode gap	0.035 in (0.90 mm)

Torque wrench settings
	lbf ft	Nm
Spark plugs	7	10
Distributor flange nuts	19	25

1   General description

In order that the engine can run correctly it is necessary for an electrical spark to ignite the fuel/air mixture in the combustion chamber at exactly the right moment in relation to engine speed and load. The ignition system is based on feeding low tension (LT) voltage from the battery to the coil where it is converted to high tension (HT) voltage. The high tension voltage is powerful enough to jump the spark plug gaps in the cylinders many times a second under high compression pressures, providing that the system is in good condition and that all adjustments are correct.

The ignition system is divided into two circuits; the low tension circuit and the high tension circuit.

The low tension (sometimes known as the primary) circuit consists of the battery lead to the starter solenoid, lead to the ignition switch, lead from the ignition switch to the low tension or primary coil windings (terminal +), and the lead from the low tension coil windings (coil terminal -) to the contact breaker points and condenser in the distributor.

The high tension circuit consists of the high tension or secondary coil windings, the heavy ignition lead from the centre of the coil to the centre of the distributor cap, the rotor arm, and the spark plug leads and spark plugs.

The system functions in the following manner. Low tension voltage is changed in the coil into high tension voltage by the opening and closing of the contact breaker points in the low tension circuit. High tension voltage is then fed via the carbon brush in the centre of the distributor cap to the rotor arm of the distributor cap, and each time it comes in line with one of the metal segments in the cap, which are connected to the spark plug leads, the opening and closing of the contact breaker points causes the high tension voltage to build up, jump the gap from the rotor arm to the appropriate metal segment and so via the spark plug lead to the spark plug, where it finally jumps the spark plug gap before going to earth.

The ignition is advanced and retarded automatically, to ensure the spark occurs at just the right instant for the particular load at the prevailing engine speed.

The ignition advance is controlled both mechanically and by a vacuum operated system. The mechanical mechanism consists of two weights, which move out from the distributor shaft, and so advance the spark. The weights are held in position by two light springs and it is the tension of the springs which is largely responsible for correct spark advancement.

The vacuum control consists of a diaphragm, one side of which is connected via a small bore tube to the carburettor, and the other side to the contact breaker plate. Depression in the inlet manifold and carburettor, which varies with engine speed and throttle opening causes the diaphragm to move, so moving the contact breaker plate, and advancing or retarding the spark. A fine degree of control is achieved by a spring in the vacuum assembly.

2   Contact breaker gap – adjustment

1   Extract the two distributor cap securing screws, pull off the cap and move it to one side without detaching the high tension leads from it. Remove the rotor arm (photos).
2   Turn the crankshaft until the heel of the breaker arm is on the high point of a cam lobe. The crankshaft can either be turned by applying a ring spanner to the crankshaft pulley bolt or by engaging top gear and pushing the car forward. Either method will be facilitated if the spark plugs are first removed.
3   Using feeler blades, check the gap is according to specification. If the points have been in use for a considerable mileage, do not make a false assessment of the gap as there will probably be a 'pip' on one face which will make precise checking impossible. In this case, the points should be dressed or renewed as described in the following Section.
4   If adjustment is required, slacken the contact set securing screw and insert a screwdriver in the slot provided and gently move the breaker arm until the clearance is correct. Retighten the securing screw (photo).
5   Before fitting the rotor arm, apply a few drops of oil to the felt pad in the centre of the cam and also squirt a few drops into the two holes in the baseplate to lubricate the mechanical advance mechanism.

2.1A Extracting distributor cap screw

2.1B Removing rotor

2.4 Checking contact points gap

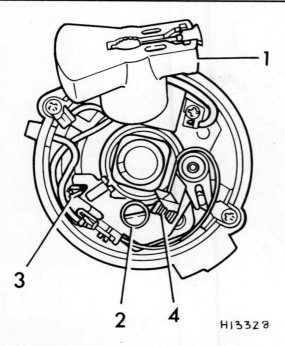

Fig. 4.1 Distributor contact breaker (Sec 2)

1   Rotor arm   3   Adjustment notch
2   Fixed arm securing screw   4   Contact points

6   Apply a little high melting point grease to the high points of the cam. If a cam wiping pad is fitted, do not apply oil to it.
7   Refit the rotor arm, wipe out the interior of the distributor cap and examine the central carbon brush in the cap before refitting it.
8   With modern engines, setting the contact breaker gap with feeler blades should be regarded as the initial adjustment only and a dwell meter should then be attached and the dwell angle checked and adjusted as described in Section 4.

3   Contact breaker points – renewal

1   Remove the distributor cap and rotor arm.
2   Remove the contact set securing screw together with the spring and flat washers.
3   Press the spring arm of the movable breaker arm towards the cam and release it from the terminal plate. Remove the contact breaker assembly.
4   If the points are not badly pitted or corroded, draw a piece of fine 'wet and dry' paper between them to clean them and then refit them. If they are badly worn or pitted, renew them. Severely pitted points may be due to a poor earth (battery or engine earth strap), a faulty condenser due to poor earth connection inside the distributor (condenser securing screw or baseplate earth wire).
5   Before fitting new points, wipe their faces clean with methylated spirit and lightly grease the pivot post.
6   After fitting the contact set, adjust the gap, as described in the preceding Section.
7   Apply a smear of high melting point grease to the high points of the cam. If a cam wiping pad is fitted, do not apply oil to it.
8   Refit the rotor arm and distributor cap.
9   Check and adjust the dwell angle as described in the next Section.

4   Dwell angle – checking and adjustment

1   For optimum engine performance, the dwell angle should always be checked after the contact breaker points have been cleaned or renewed.

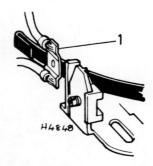

Fig. 4.2 Movable contact arm and low tension terminal plate (1)
(Sec 3)

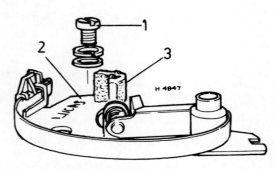

Fig. 4.3 Contact breaker set (Sec 3)

1   Fixed arm securing screw   3   Cam wiping pad
2   Fixed contact arm

2   The dwell angle is the number of degrees through which the distributor cam turns during the period between the instants of closure and opening of the contact breaker points. It can only be checked with a dwell meter with the engine running, the meter usually being connected between the coil negative terminal and a good earth.
3   If the dwell angle is outside that specified, switch off the engine and adjust the points in the following way:

   If the dwell angle is too large, increase the points gap
   If the dwell angle is too small, reduce the points gap

4   Only make very fine adjustments and re-check the dwell angle after each alteration until it is within the permitted tolerance.
5   The dwell angle should always be adjusted before carrying out ignition timing.
6   Always check the ignition timing after altering the dwell angle.

5   Ignition timing

1   Two instruments will be required for this work, a stroboscope and a tachometer. Connect these to the engine in accordance with the maker's instructions. Have the engine at normal operating temperature.
2   Mark the appropriate BTDC mark on the crankshaft pulley and the end of the timing pointer with quick drying white paint (refer to Specifications for specified advance).
3   Disconnect the vacuum pipe from the carburettor and plug the pipe.
4   Start the engine and allow it to run at the specified speed:

   Single carburettor engine   *1500 rev/min*
   Twin carburettor engine   *2000 rev/min*

5   Point the stroboscope at the white marks when they will appear stationary. If the timing is correct, they will also appear in alignment.

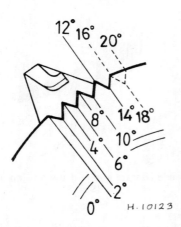

Fig. 4.4 Timing marks aligned at 8° BTDC (Sec 5)

6.4 Extracting condenser screw

If the marks do not appear to be in alignment, release the distributor mounting flange nuts and turn the distributor body fractionally in either direction until the timing marks coincide.

6   Retighten the nuts, switch off the engine and remake the original connections.

7   The engine is designed to have the ignition timing checked using LED (light emitting diode) optical sensor.

8   This will usually only be carried out by dealers having suitable equipment, but the home mechanic will wish to know that the timing disc located behind the crankshaft pulley and the sensor bracket on the front of the engine are essential parts of this system.

6 Condenser (capacitor) – removal, testing and refitting

1   The condenser ensures that with the contact breaker points open, the sparking between them does not cause severe pitting. The condenser is fitted in parallel and its failure will automatically cause failure of the ignition system as the points will be prevented from interrupting the low tension circuit.

2   Testing for an unserviceable condenser may be effected by switching on the ignition and separating the contact points by hand. If this action is accompanied by a strong blue flash then condenser failure is indicated. Difficult starting, missing of the engine after several miles running or badly pitted points are other indications of a faulty condenser.

3   The surest test is by substitution of a new unit.

4   Removal of the condenser is by means of withdrawing the screw which retains it to the distributor baseplate. Replacement is a reversal of this procedure (photo).

7 Distributor – removal and refitting

1   Extract the screws and pull off the distributor cap.

2   Disconnect the low tension wire from the negative terminal on the ignition coil.

3   Disconnect the vacuum pipe from the distributor (photo).

4   Remove No 1 spark plug and turn the crankshaft by means of the crankshaft pulley bolt until compression can be felt if the thumb is placed over the plug hole, which will indicate that the piston is rising on its compression stroke.

5   Continue turning until the 8° BTDC mark on the crankshaft pulley is opposite the pointer.

6   The rotor arm will now be positioned as shown in Fig. 4.6.

7   Mark the rim of the distributor body opposite to the contact end of the rotor arm.

8   Mark the position of the studs within the mounting flange slots and then remove these two nuts which hold the distributor flange to the crankshaft cover.

9   Withdraw the distributor.

10 Before fitting the distributor, renew the O-ring seal.

7.3 Vacuum pipe at distributor

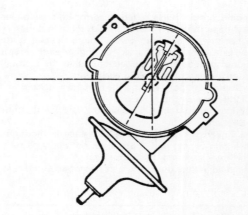

Fig. 4.5 Rotor arm position with engine set at 8° BTDC, before installing distributor (Sec 7)

Are your plugs trying to tell you something?

Normal.
Grey-brown deposits, lightly coated core nose. Plugs ideally suited to engine, and engine in good condition.

Heavy Deposits.
A build up of crusty deposits, light-grey sandy colour in appearance.
Fault: Often caused by worn valve guides, excessive use of upper cylinder lubricant, or idling for long periods.

Lead Glazing.
Plug insulator firing tip appears yellow or green/yellow and shiny in appearance.
Fault: Often caused by incorrect carburation, excessive idling followed by sharp acceleration. Also check ignition timing.

Carbon fouling.
Dry, black, sooty deposits.
Fault: over-rich fuel mixture.
Check: carburettor mixture settings, float level, choke operation, air filter.

Oil fouling.
Wet, oily deposits. Fault: worn bores/piston rings or valve guides; sometimes occurs (temporarily) during running-in period.

Overheating.
Electrodes have glazed appearance, core nose very white – few deposits. Fault: plug overheating. Check: plug value, ignition timing, fuel octane rating (too low) and fuel mixture (too weak).

Electrode damage.
Electrodes burned away; core nose has burned, glazed appearance. Fault: pre-ignition. Check: for correct heat range and as for 'overheating'.

Split core nose.
(May appear initially as a crack). Fault: detonation or wrong gap-setting technique. Check: ignition timing, cooling system, fuel mixture (too weak).

WHY DOUBLE COPPER IS BETTER FOR YOUR ENGINE.

Unique Trapezoidal Copper Cored Earth Electrode — 50% Larger Spark Area — Copper Cored Centre Electrode

Champion Double Copper plugs are the first in the world to have copper core in both centre and earth electrode. This innovative design means that they run cooler by up to 100°C – giving greater efficiency and longer life. These double copper cores transfer heat away from the tip of the plug faster and more efficiently. Therefore, Double Copper runs at cooler temperatures than conventional plugs giving improved acceleration response and high speed performance with no fear of pre-ignition.

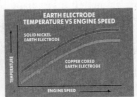

Champion Double Copper plugs also feature a unique trapezoidal earth electrode giving a 50% increase in spark area. This, together with the double copper cores, offers greatly reduced electrode wear, so the spark stays stronger for longer.

 FASTER COLD STARTING

 FOR UNLEADED OR LEADED FUEL

 ELECTRODES UP TO 100°C COOLER

 BETTER ACCELERATION RESPONSE

 LOWER EMISSIONS

 50% BIGGER SPARK AREA

 THE LONGER LIFE PLUG

Plug Tips/Hot and Cold.

Spark plugs must operate within well-defined temperature limits to avoid cold fouling at one extreme and overheating at the other.

Champion and the car manufacturers work out the best plugs for an engine to give optimum performance under all conditions, from freezing cold starts to sustained high speed motorway cruising.

Plugs are often referred to as hot or cold. With Champion, the higher the number on its body, the hotter the plug, and the lower the number the cooler the plug. For the correct plug for your car refer to the specifications at the beginning of this chapter.

Plug Cleaning

Modern plug design and materials mean that Champion no longer recommends periodic plug cleaning. Certainly don't clean your plugs with a wire brush as this can cause metal conductive paths across the nose of the insulator so impairing its performance and resulting in loss of acceleration and reduced m.p.g.

However, if plugs are removed, always carefully clean the area where the plug seats in the cylinder head as grit and dirt can sometimes cause gas leakage.

Also wipe any traces of oil or grease from plug leads as this may lead to arcing.

CHAMPION

DOUBLE COPPER

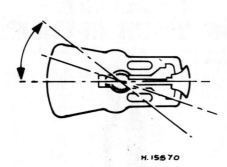

Fig. 4.6 Rotor position after installing distributor (Sec 7)

11 Check that the engine is still timed to 8° BTDC. If it is not, set it as described earlier in this Section.

12 Hold the distributor in position so that the vacuum unit is pointing downwards to the left at an angle of 45° to the surface of the cam cover (Fig. 4.5).

13 Turn the rotor arm until it is at the one o'clock position.

14 Push the distributor into position. As the gears mesh, the rotor should turn until it takes up the position which it held before removal (parallel to cam cover and aligned with mark made on distributor rim) (Fig. 4.6).

15 Turn the distributor until the points are just about to open and tighten the flange screws.

16 Fit the distributor cap, connect the low tension lead and the vacuum pipe and then check the dwell angle and timing (see Sections 4 and 5).

8 Distributor – overhaul

1 Before overhauling a well worn distributor, consideration should be given to obtaining a factory reconditioned unit. If it is decided to dismantle the original unit, first check the availability of spare parts.

2 With the distributor removed as previously described in this Chapter, withdraw the cap and rotor arm and extract the felt lubrication pad from the centre of the cam.

3 Remove the two vacuum unit retaining screws, tilt the unit to disengage the operating arm and then withdraw the vacuum unit.

4 Push the low tension lead and grommet into the interior of the distributor body (photo).

5 Extract the contact breaker screw and remove the contact breaker assembly (photo).

6 Remove the condenser and earth lead.

7 Extract the screws and lift out the baseplate. Note that one of the baseplate screws also holds the condenser earth lead (photos).

8 Note carefully the type and position of the counterweight springs and detach them (photo).

9 If the drivegear is worn and must be dismantled, drive out the roll pins from the drivegear.

10 Remove the drivegear and the thrust washer.

11 Withdraw the distributor shaft complete with centrifugal advance mechanism, spacer and washer.

12 Inspect all components for wear. If the contacts in the cap or the contact end of the rotor are burned, do not attempt to clean them but renew the components.

13 Do not dismantle the centrifugal advance mechanism beyond detaching the springs.

14 If the distributor shaft has any side movement due to wear, do not attempt to remove the bearing but renew the complete distributor, if you are quite sure that it is the bearing that is worn and not the shaft.

15 Commence reassembly with the drivegear and shaft. Make sure that the spacer is fitted with its chamfer towards the stud washer.

16 If the original shaft is being refitted, fit the thrust washer and drivegear, align the rotor arm keyway with the hole in the drivegear and tap in the roll pin.

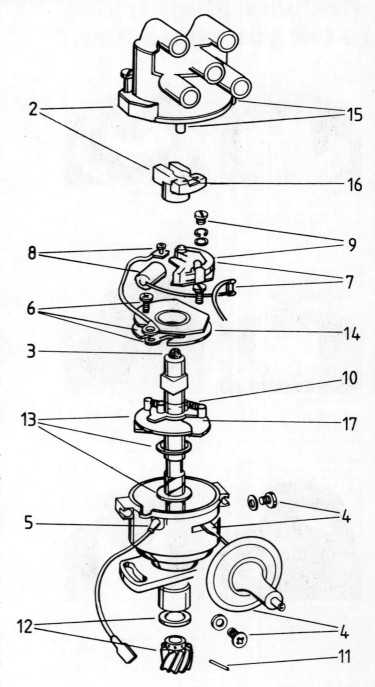

Fig. 4.7 Exploded view of distributor (Sec 8)

2 *Cap*	11 *Roll pin*
3 *Felt oil pad*	12 *Drivegear and thrust washer*
4 *Vacuum unit screws*	13 *Centrifugal advance*
5 *Low tension lead and*	*mechanism, spacer and*
grommet	*steel washer*
6 *Baseplate and screws*	14 *Fixed baseplate*
7 *Low tension connector and*	15 *Carbon brush in*
spring contact arm	*distributor cap*
8 *Condenser and earth lead*	16 *Rotor*
9 *Contact set screw*	17 *Counterweight*
10 *Counterweight springs*	

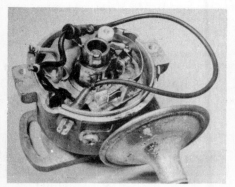
8.4 Distributor LT lead

8.5 Removing contact breaker

8.7A Extracting baseplate screw

8.7B Removing baseplate

8.8 Distributor centrifugal advance mechanism

8.19 Lubricating centrifugal advance mechanism

17 If a new shaft is being fitted, fit the thrust washer, drivegear and the rotor arm to the shaft. Set the rotor arm at 90° to a line drawn through the centres of the two mounting flange bolt holes at the base of the distributor so that the non-contact end of the rotor arm is towards the thrust washer tab slot in the distributor body. Insert a 0.005 in (0.13 mm) feeler blade between the thrust washer and drivegear so that when it is eventually withdrawn the correct shaft endfloat will be established. Align the hole in the drivegear with the edge of the tab slot. Hold the shaft and drivegear tightly together and drill through the shaft using a 3.2 mm diameter drill. Remove the feeler blade and drive in the roll pin.

18 The remainder of the reassembly operations are a reversal of dismantling, but check that the operating arm has engaged with the advance plate post before screwing in the securing screws.

19 Apply a little oil to the advance mechanism (photo).

9 Ignition coil

1 It is important that the low tension lead from the distributor always connects to the negative (-) terminal on the coil.

2 Occasionally wipe the terminal end of the coil clean and check that the coil mounting clip is tight.

3 Without special equipment, the best method of testing for a faulty coil is by substitution of a new unit. Before doing this however, check the original coil for security of the connecting leads and remove any corrosion which may have built up in the coil high tension lead socket.

10 Spark plugs and high tension leads

1 The correct functioning of the spark plugs is vital for the correct running and efficiency of the engine. The plugs fitted as standard are listed on the Specifications page.

2 Clean and regap the spark plugs at specified intervals and renew them also as specified in Routine Maintenance. The condition of the spark plug will also tell much about the overall condition of the engine.

3 If the insulator nose of the spark plug is clean and white, with no deposits, this is indicative of a weak mixture, or too hot a plug. (A hot plug transfers heat away from the electrode slowly — a cold plug transfers it away quickly).

4 If the top and insulator nose is covered with hard black looking deposits, then this is indicative that the mixture is too rich. Should the plug be black and oily, then it is likely that the engine is fairly worn, as well as the mixture being too rich.

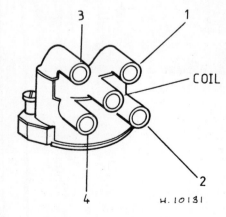

Fig. 4.8 Spark plug lead connections at distributor cap (Sec 10)

5   If the insulator nose is covered with light tan to greyish brown deposits, then the mixture is correct and it is likely that the engine is in good condition.

6   If there are any traces of long brown tapering stains on the outside of the white portion of the plug, then the plug will have to be renewed as this shows that there is a faulty joint between the plug body and the insulator, and compression is being allowed to leak away.

7   Plugs should be cleaned by a sand blasting machine, which will free them from carbon more thoroughly than cleaning with a wire brush. The machine will also test the condition of the plugs under compression. Any plug that fails to spark at the recommended pressure should be renewed.

8   The spark plug gap is of considerable importance, as, if it is too large or too small the size of the spark and its efficiency will be seriously impaired. The spark plug gap should be set to 0.035 in (0.90 mm) for the best results.

9   To set it, measure the gap with a feeler gauge, and then bend open or close, the outer plug electrode until the correct gap is achieved. The centre electrode should never be bent as this may crack the insulation and cause plug failure, if nothing worse.

10   The spark plugs do not have sealing washers but rely on the contact of the conical faces of plug and seat, which should always be wiped perfectly clean before fitting.

11   Never overtighten a spark plug, use a torque wrench where possible to tighten to the specified figure (photo).

12   Connect the high tension leads to the spark plugs in the correct firing order, No 1 plug being the one nearest the crankshaft pulley.

13   The spark plug leads require no attention other than being kept securely attached and wiped clean at regular intervals.

14   When renewing spark plug leads, always renew them with ones of similar suppressed type.

---

11   Ignition/starter switch – removal and refitting

1   Extract the screws and pull away the steering column upper shrouds. On single carburettor models there is no need to disconnect the choke control (photo).

2   Unbolt the base of the steering column from the pedal bracket.

3   Unscrew the pinch-bolt and release the upper end of the steering column. Lower the steering column carefully.

4   Extract the small screw and withdraw the ignition/starter switch from the steering lock (photo).

5   Unplug the wiring harness connector and remove the switch.

6   Refitting is a reversal of removal.

7   Removal of the steering lock is described in Chapter 11.

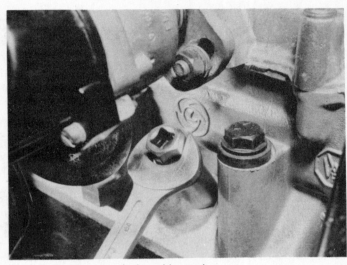

10.11 Tightening a spark plug with a socket

11.1 Ignition key markings

11.4 Removing ignition switch

12 Fault diagnosis – ignition system

Symptom	Reason(s)
Engine fails to start	Discharged battery Oil on points Loose battery connections Disconnected leads Faulty condenser Damp leads or distributor cap
Engine misfires	Faulty spark plug Cracked distributor cap Cracked rotor arm Worn advance mechanism Incorrect spark plug gap Incorrect points gap Faulty condenser Faulty coil Incorrect ignition timing Poor earth connections
Engine overheats or lacks power	Seized centrifugal weights Perforated vacuum pipe Incorrect ignition timing
Engine 'pinks'	Timing too advanced Advance mechanism stuck in advanced position Broken counterweight spring Low octane rating of fuel Excessive upper cylinder oil used in fuel Excessive oil vapour from crankcase ventilation system or worn piston rings

Chapter 5 Clutch

Contents

Specifications

System type .... Single dry plate, diaphragm spring, hydraulic operation

Hydraulic fluid

Type/specification .... Hydraulic fluid to FMVSS DOT 3 (Duckhams Universal Brake and Clutch Fluid)

Driven plate diameter .... 8.5 in (215.0 mm)

Master cylinder diameter .... 0.625 in (15.5 mm)

Slave cylinder diameter .... 1.125 in (28.57 mm)

Number of torsion springs 6

Release bearing .... Sealed ball bearing

Torque wrench settings	lbf ft	Nm
Clutch cover bolts	17	23
Master cylinder mounting nuts	16	22
Slave cylinder mounting bolts	30	41

1 General description and maintenance

1 The clutch mechanism is of diaphragm spring type with a single driven plate. Actuation is hydraulic.

2 At the specified intervals, check the fluid level in the master cylinder reservoir and if necessary, top-up to the bottom of the filler neck using only recommended fluid. Keep the vent hole in the filler cap clear (photo).

3 No adjustment is required throughout the life of the driven plate.

2 Master cylinder – removal, overhaul and refitting

1 Attach a bleed tube to the bleed nipple on the clutch slave cylinder and open the nipple. Remove the cap from the fluid reservoir and then pump the clutch pedal until all fluid has been expelled into a suitable container.

2 Disconnect the master cylinder pushrod from the clutch pedal.

3 Disconnect the hydraulic line from the master cylinder body.

4 Unscrew and remove the master cylinder flange nuts and withdraw the cylinder. The securing studs are mounted on a separate plate.

5 Clean the external surfaces of the assembly and detach the rubber dust-excluding boot.

6 Extract the circlip and withdraw the pushrod complete with dished washer.

7 Withdraw the piston assembly. Clean in methylated spirit or hydraulic fluid. Use nothing else. Examine the surfaces of the piston and cylinder. If they are scored or show any 'bright' wear areas, the master cylinder must be renewed as a complete unit.

8 Where the components are in good order, remove and discard the

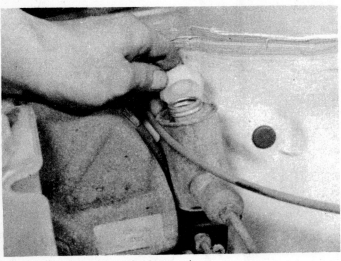

1.2 Clutch master cylinder fluid reservoir

Fig. 5.1 Clutch master cylinder pushrod attachment to pedal arm (Sec 2)

2   Spring clip and washer

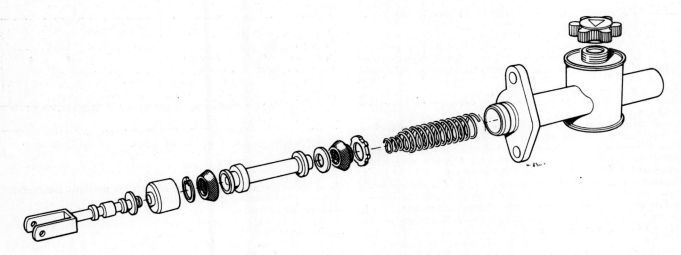

Fig. 5.2 Exploded view of clutch master cylinder (Sec 2)

rubber seals and obtain a repair kit. The repair kit will contain all the necessary new seals and other renewable items.

9   Reassembly is a reversal of dismantling but manipulate the seals using the fingers only and dip each component in clean brake fluid before assembling it.

10 Refitting is a reversal of removal but bleed the system, as described in Section 4.

3   Clutch slave cylinder – removal, overhaul and refitting

1   Drain the system by attaching a bleed tube to the slave cylinder bleed nipple and after removing the reservoir cap, pumping the clutch pedal to expel the fluid into a suitable container.

2   Disconnect the flexible hose by unscrewing the union locknut at the support bracket.

3   Unscrew and remove the two setscrews which secure the slave cylinder to the flywheel housing, withdraw the cylinder leaving the pushrod attached to the clutch release lever (photo).

4   Clean the external surfaces of the cylinder, release the retaining ring and remove the dust cover.

5   Extract the piston retaining circlip.

6   Extract the piston assembly either by tapping the cylinder body on

3.3 Unscrewing clutch slave cylinder

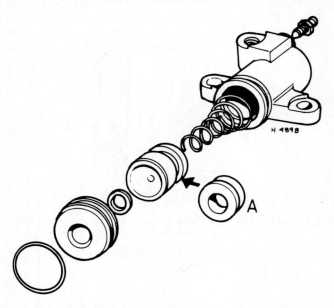

Fig. 5.3 Exploded view of clutch slave cylinder (Sec 3)

A   Piston seal

a piece of wood, or by applying air from a tyre pump at the fluid inlet port.

7   Remove the return spring and bleed screw.

8   Clean all components in methylated spirit or hydraulic fluid. Use nothing else.

9   Check the piston and cylinder bore surfaces for scoring or 'bright' wear areas. If these are evident, renew the cylinder complete.

10   If the components are in good order, remove the rubber components and discard them and obtain a repair kit. This will contain the new seals and all the other renewable components.

11   Reassembly is a reversal of dismantling but dip each part in clean hydraulic fluid before assembling and use only the fingers to manipulate the new seals in position. Note that the piston seal is fitted with its lip toward the small land of the piston and the return spring has its small end towards the piston.

12   Refitting is a reversal of removal but bleed the system, as described in Section 4.

4   Hydraulic system – bleeding

1   Fill the clutch master cylinder reservoir with clean hydraulic fluid which has been stored in an airtight container and has remained unshaken for the preceding 24 hours.

2   Attach a bleed tube to the bleed screw on the slave cylinder and immerse the open end of the tube in a jar containing an inch or two of hydraulic fluid.

3   Have an assistant depress the clutch pedal fully, then tighten the bleed screw and allow the pedal to return unassisted. Repeat the operation until no air bubbles are seen being expelled from the end of the tube in the jar.

4   During the bleeding process, maintain the fluid level in the reservoir at least half-full and then finally top-up to the full mark when the operations are complete.

5   The clutch hydraulic system may be bled using a one way valve or pressure bleeding kit. Refer to Chapter 9 where these systems and their operation are described.

5   Clutch pedal – removal and refitting

1   If a clutch pedal return spring is fitted, disconnect it.

2   Disconnect the master cylinder pushrod from the clutch pedal.

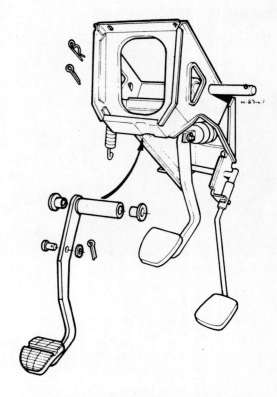

Fig. 5.4 Control pedal arrangement (Sec 5)

3   Extract the split pin from the left-hand end of the pedal shaft.

4   Unscrew and remove the self-tapping screw which holds the spring pin to the bracket. Remove the spring pin from the left-hand end of the shaft.

5   Remove the upper outer insulation pad from the bulkhead. On cars with left-hand steering, pull the insulation back to the right-hand side of the pedal bracket.

6   Pull the pedal cross-shaft out of the pedal arms and remove the clutch pedal.

7   Examine the pedal pivot bushes for wear. Renew them if necessary according to type, bronze or nylon.Smear them with grease on assembly.

8   Refitting is a reversal of removal.

6   Clutch – removal

1   Refer to Chapter 6 and remove the flywheel housing.

2   Unscrew the clutch cover bolts evenly, a turn at a time, until the diaphragm spring pressure is relieved. Mark the clutch cover in relation to the flywheel.

3   Pull the clutch cover from its flywheel dowels and catch the driven plate as it is released.

7   Clutch components – inspection

1   Owing to the slow-wearing qualities of the clutch, it is not easy to decide when to go to the trouble of removing it in order to check the wear on the friction lining. The only positive indication that something needs doing is when it starts to slip or when squealing noises on engagement indicate that the friction lining has worn down to the rivets. In such instances it can only be hoped that the friction surfaces on the flywheel and pressure plate have not been badly worn or scored.

2   A clutch will wear according to the way in which it is used. Much intentional slipping of the clutch while driving – rather than the correct

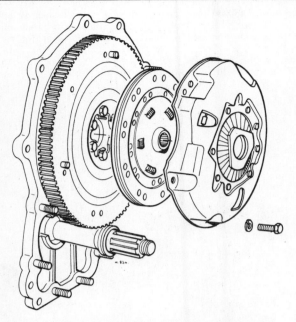

Fig. 5.5 Clutch driven plate and cover (Sec 6)

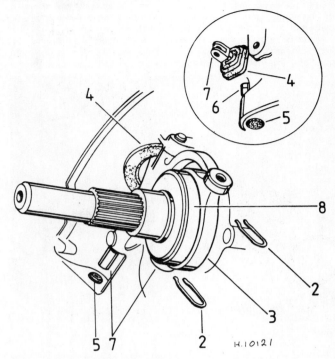

Fig. 5.6 Clutch release components (Sec 8)

2	*Spring clips*	6	*Lock screw*
3	*Release bearing carrier*	7	*Pivot shaft and clevis*
4	*Boot*		*fork*
5	*Grommet*	8	*Release bearing*

selection of gears – will accelerate wear. It is best to assume however, that the driven plate may need renewal any time after 30 000 miles (48 000 km) have been covered.

3   Examine the surfaces of the pressure plate and flywheel for signs of scoring. If this is only light it may be left, but if very deep the pressure plate unit will have to be renewed. If the flywheel is deeply scored it should be taken off and advice sought from an engineering firm. Providing it may be machined completely across the face the overall balance of engine and flywheel should not be too severely upset. If renewal of the flywheel is necessary the new one will have to be balanced to match the original.

4   If the surface of the flywheel is covered with tiny cracks, these are caused by overheating and it may be possible to machine them out, otherwise a new flywheel will be required.

5   Driven plate lining surfaces should be at least $\frac{1}{32}$ in (0.8 mm) above the rivets, otherwise the disc is not worth putting back. If the lining material shows signs of breaking up or black areas where oil contamination has occurred it should be renewed.

6   Examine the tips of the fingers of the diaphragm spring. If these are worn or stepped as a result of contact with the release bearing then the pressure plate assembly (clutch cover) must be renewed.

8   Clutch release bearing – renewal

1   Renewal of the clutch release bearing will normally be carried out at the same time as the clutch is dismantled. At other times, the bearing is accessible after removing the flywheel housing (see Chapter 6).

2   Prise the spring clips from both sides of the release bearing (photo).

3   Pull the release bearing assembly off the clutch shaft (photo).

4   If further dismantling of the release mechanism is required, remove the withdrawal lever boot and the rubber grommet from the flywheel housing.

5   Remove the pivot shaft locking washer.

6   Withdraw the pivot shaft and take off the clutch withdrawal fork.

7   Reassembly and refitting are reversals of removal and dismantling.

9   Clutch – refitting

1   With clean hands, locate the new driven plate on the flywheel so that the larger hub of the plate (marked FLYWHEEL SIDE) is towards the flywheel (photo).

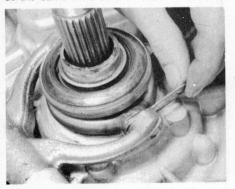

8.2 Clutch release bearing retaining spring

8.3 Removing clutch release bearing

9.1 Driven plate marking

2   Offer the cover onto its locating dowels and screw in the retaining bolts just finger tight. The cover is coated with a protective coating, do not attempt to remove this with solvent (photo).

3   A rod or stepped mandrel must now be used to centralise the driven plate before the clutch cover bolts are tightened more than finger-tight. The centralising tool must be able to pass through the splined hub of the driven plate and engage in the crankshaft spigot bush. As the tool is inserted the driven plate will be centralised and the cover bolts can then be tightened to their final specified torque (photos).

4   Remove the centralising tool.

5   Fit the flywheel housing by reference to Chapter 6.

9.2 Fitting clutch driven plate and cover

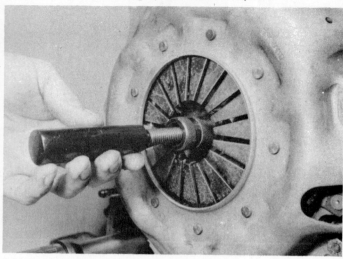

9.3A Centralising clutch driven plate

9.3B Tightening clutch cover bolts

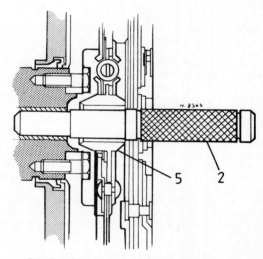

Fig. 5.7 Clutch alignment tool (Sec 9)

2   Tool
5   Driven plate splined
     hub

10 Fault diagnosis – clutch

Symptom	Reason(s)
Judder when taking up drive	Loose engine or gearbox mountings Badly worn friction surfaces or contaminated with oil Worn splines on gearbox input shaft or driven plate hub Worn input shaft spigot bush in flywheel
*Clutch spin (failure to disengage) so that gears cannot be meshed	Driven plate sticking on input shaft splines due to rust. May occur after vehicle standing idle for long period Damaged or misaligned pressure plate assembly
Clutch slip (increase in engine speed does not result in increase in vehicle road speed – particularly on gradients)	Incorrect release bearing to diaphragm spring finger clearance Friction surfaces worn out or oil contaminated
Noise evident on depressing clutch pedal	Dry, worn or damaged release bearing Weak or broken pedal return spring Weak or broken clutch release lever return spring Excessive play between driven plate hub splines and input shaft splines
Noise evident as clutch pedal released	Distorted driven plate Broken or weak driven plate cushion coil springs Weak or broken clutch pedal return spring Weak or broken release lever return spring Distorted or worn input shaft Release bearing loose on retainer hub

This condition may also be due to the driven plate being rusted to the flywheel or pressure plate. It may be possible to free it by applying the handbrake, engaging top gear, depressing the clutch pedal and operating the starter motor. If really badly corroded, then the engine will not turn over, but in the majority of cases the driven plate will free. Once the engine starts, increase its speed and slip the clutch several times to clear the rust deposits.

Chapter 6 Manual gearbox and automatic transmission

Contents

Specifications

Part 1 Manual gearbox

Type .... 4-speed (synchromesh) with reverse. Transversely mounted with final drive housing attached at base of engine crankcase. Floor mounted gearchange.

Ratios
1st .... 3.545 : 1
2nd .... 2.217 : 1
3rd .... 1.439 : 1
4th .... 1.000 : 1
5th .... 3.312 : 1

Mainshaft endfloat .... 0.006 to 0.008 in (0.15 to 0.20 mm)

3rd speed gear endfloat .... 0.006 to 0.008 in (0.15 to 0.20 mm)

2nd speed gear endfloat .... 0.005 to 0.008 in (0.13 to 0.20 mm)

Oil capacity .... Combined with engine – refer to engine specifications

Part 2 Automatic transmission

Type .... Borg-Warner Type 35TA three-speed with torque converter and epicyclic gear train. Mounted transversely at base of engine crankcase with final drive housing attached.

Ratios
1st .... 2.39 : 1
2nd .... 1.45 : 1
3rd .... 1.00 : 1
Reverse .... 2.09 : 1

Lubrication
Lubricant type/specification .... Automatic transmission fluid to M2C 33G (Duckhams Q-Matic)
Capacity:
    From dry .... 13.0 pints (7.4 litres)
    At drain and refill .... 8.0 pints (4.5 litres)
    Quantity retained in torque converter .... 5.0 pints (3.0 litres)
    Fluid cooler capacity .... 1.75 pints (1.0 litre)

Torque wrench settings

	lbf ft	Nm
Manual gearbox		
Drain plug	20	27
Flywheel housing to adaptor:		
Smaller bolts	20	27
Larger bolts	35	47
Front cover bolts	18	24
Gearbox to crankcase:		
$\frac{5}{16}$ in UNF	25	34
$\frac{5}{16}$ in UNC	12	16
$\frac{3}{8}$ in UNF	30	41
Selector rod retaining plate screw	16	22
Input shaft gear nut	120	163
Input shaft bearing nut	120	163
Mainshaft nut	40	54
Mainshaft bearing housing screw	18	24
Mainshaft bearing locating plate screw	15	20
Adaptor plate to engine	30	41
Clutch shaft nut	60	81
Clutch shaft bearing retainer screws	15	20
Mainshaft pinion nut	150	204
Automatic transmission		
Chain cover to torque converter housing	9	12
Torque converter shaft nut	70	95
Input shaft nut	30	41
Driveplate to torque converter	30	41
Servo cover bolts	8	11
Crankcase to transmission casing	30	41
Pinion retaining nut	160	218
Fluid restrainer cover	20	27
Converter housing to transmission case	10	14
Driveplate to crankshaft	60	82
Valve body to converter housing	8	11

PART 1: MANUAL GEARBOX

General description and maintenance

1 The manual gearbox is of four-speed synchromesh type.
2 The final drive/differential is attached to the gearbox and the complete transmission unit is mounted transversely at the base of the crankcase.
3 A common oil supply provides lubrication for the engine, gearbox and final drive. Refer to Chapter 1, Section 2 for details of topping up and draining and refilling.
4 Gear selection is by means of a floor-mounted lever and remote control rod.

2 Operations possible without removing gearbox

The following operations are possible without removing the gearbox:

(a) Front cover – removal and refitting
(b) Gearchange mechanism – removal, overhaul and refitting
(c) Gear selector mechanism – removal, overhaul and refitting
(d) Flywheel housing – removal and refitting

3 Front cover – removal and refitting

1 Drain the engine/transmission oil and then refit the plug.

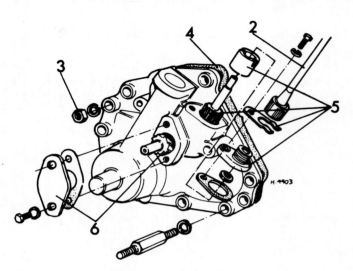

Fig. 6.1 Front cover (Sec 3)

2   *Speedometer cable*
3   *Nut*
4   *Gasket*

5   *Speedometer driven gear,
    bushes and fork*
6   *Speedometer drivegear
    and endplate*

2   Disconnect the speedometer drive cable from the transmission.
3   Unscrew the pillar stud and nuts and remove the front cover and gasket.
4   The oil pick-up/filter can be dismantled by removing the fixing nut and washers, pulling off the casing, extracting the seal, discs and coil spring (photos).

5   The speedometer pinion can be removed after unbolting the forked clamp plate.
6   The speedometer drivegear can be removed if the diamond shaped plate is unbolted.
7   Refitting is a reversal of removal. Use a new gasket and refill the engine/transmission with oil.

4   Gearchange lever and remote control rods – removal, overhaul and refitting

1   Unscrew and remove the gear lever knob.
2   Raise the front of the car and support it under the right-hand front side jacking point.
3   From below the car, peel back the draught excluder and release the gear lever bayonet type securing cap.
4   Lift the gear lever assembly up into the draught excluder so that it clears the remote control mechanism and then extract the gearchange lever assembly from below the car. Detach the bayonet cap from the lever.
5   To remove the remote control assembly, drive out the roll pin which retains the extension rod to the selector rod at the gearbox selector housing end.
6   Remove the bolt which secures the remote control steady rod to the selector housing.
7   Disconnect the leads from the reversing lamp switch.
8   Remove the bolt which secures the remote control assembly to its mounting brackets and then withdraw the assembly downwards.
9   Secure the remote control assembly in a vice and remove the bottom cover and reverse light plates.
10 Remove the steady rod from the housing.
11 Remove the reverse light switch.
12 Move the extension rod eye to the rear and then drive out the roll pin which secures the extension rod to the eye.
13 Push the extension rod eye forwards and remove the roll pin which retains the support rod to the extension rod eye. Remove the extension rod.

3.4A Transmission oil pick-up filter casing nut

3.4B Removing filter casing

3.4C Filter seal

3.4D Filter discs

3.4E Filter spring

3.4F Filter components removed from front cover

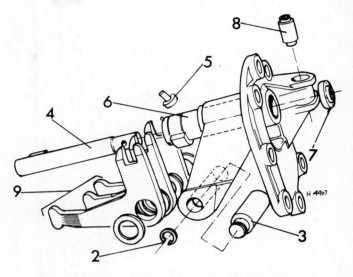

Fig. 6.3 Gear selector mechanism and housing (Sec 5)

2   Circlip   6   Interlock spool
3   Pivot pin   7   Selector shaft oil seal
4   Selector shaft   8   Mounting bush
5   Interlock spool locating pin   9   Bellcrank levers

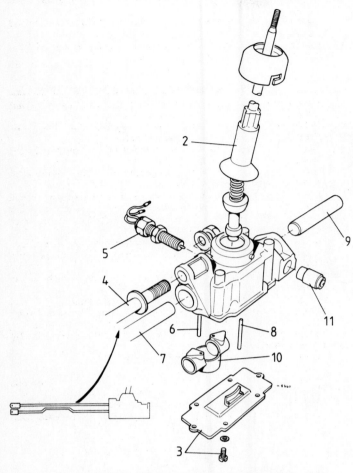

Fig. 6.2 Gearchange remote control components (Sec 4)

2   Gear lever   7   Extension rod
3   Bottom cover   8   Roll pin
4   Steady rod   9   Support rod
5   Reversing lamp switch   10 Extension rod eye
6   Roll pin   11 Mounting bushes

14 Drive out the support and then lift out the extension rod eye.
15 If the mounting bushes are worn, they can be extracted and new ones fitted.
16 Reassembly is a reversal of dismantling but apply grease liberally to all friction surfaces and ensure that the cranked section of the extension rod is uppermost so that it will clear the exhaust system. Make sure that the extension rod eye is the correct way up before connecting the rod to it.
17 Refitting is a reversal of removal.
18 Adjustment of the reversing lamp switch should be checked. If necessary, release the switch locknut, select reverse gear and then with a battery and test lamp connected between the switch terminals, screw in the switch until the lamp lights. Screw the switch in a further five hexagon flats and then tighten the locknut. Remake the original connections.

5   Gear selector mechanism – removal, overhaul and refitting

1   Drain the engine/transmission oil and refit the plug. Disconnect the exhaust downpipes and the front suspension cross-tube.
2   Drive out the roll pin which secures the gearchange extension rod to the selector shaft.

3   Remove the bolt which secures the remote control steady rod to the selector housing.
4   Remove the six nuts which secure the selector housing to the final drive housing and then pull the selector housing from its studs.
5   Extract the circlip from the selector lever pivot and remove the pivot and the shaft. Drive out the old oil seal and fit a new one.
6   Remove the circlip which secures the bellcrank lever pivot pin to the housing.
7   Withdraw the pivot pin and bellcrank levers, retaining the washers which are located between and on each side of the levers.
8   Withdraw the selector shaft.
9   Push the interlock spool locating pin into the spool and then remove the pin.
10 Renew any worn components and reassemble by reversing the dismantling procedure.
11 Refitting the housing is a reversal of removal but use a new flange gasket and make quite sure that the selector levers have engaged positively with the lugs on the forks before tightening the housing bolts.
12 Reconnect the extension and steady rods and fill the engine/transmission with oil.

6   Flywheel housing – removal and refitting

1   Remove the battery.
2   Remove the support brackets for the cooling system expansion tank and the ignition coil and move the tank and coil aside.
3   Remove the starter motor.
4   Unbolt the clutch slave cylinder and tie it up out of the way. Detach the clutch release lever return spring.
5   Raise the front of the car and support it on stands.
6   Drain the engine oil and refit the drain plug.
7   Remove the exhaust mounting nuts to the body support bracket and remove the left-hand mounting from the exhaust pipe.
8   Disconnect the gearchange linkage from the transmission casing. To do this, drive out the roll pin which holds the extension rod to the selector shaft and then remove the bolt which holds the remote control steady rod to the selector housing. Tie the gearchange rods loosely to the steering rack.
9   Release the right-hand driveshaft from the differential as described in Chapter 8.

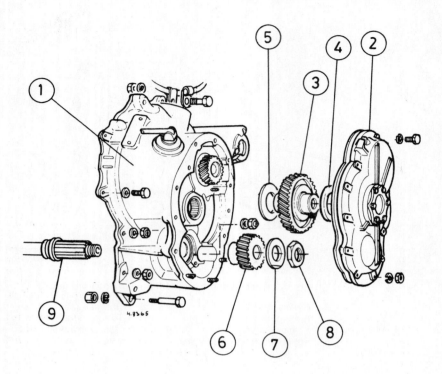

Fig. 6.4 Flywheel housing and primary drive (Sec 6)

1   Flywheel housing
2   Primary drive cover
3   Idler gear
4   Thrust washer
5   Thrust washer
6   Input shaft gear
7   Lockplate
8   Shaft nut
9   Input shaft

10   Release the bonnet support struts from their body brackets and tie the bonnet back to give room for the attachment of an engine lifting hoist later.

11   Remove the ten screws and three nuts and lift away the primary drive cover and joint washer, noting the location of the clutch release lever return spring bracket. Be prepared for some loss of oil. Manoeuvre the cover downward and remove it.

12   Mark the end of the now exposed idler gear with quick drying paint so that it can be refitted the correct way round.

13   Turn the flywheel so that the gearbox input shaft nut locking tab is at the bottom. Lock the flywheel ring gear and then unscrew the nut. Remove the lockwasher, idler gear and thrust washers. These thrust washers are fitted over each side of the idler gear with their oil grooves away from the gear.

14   Unbolt the bearing retainer and withdraw the clutch shaft and gear.

15   Remove the air cleaner and the oil filter cartridge.

16   Fit lifting eyes to the four corner studs on the cylinder head. Attach a suitable hoist and take the weight of the engine.

17   Extract the bolts which hold the front right-hand engine mounting bracket to the flywheel housing.

18   Extract the long bolts from the remaining engine mountings.

19   Remove the screw and nuts which hold the flywheel housing to the adaptor plate. Note the position of the earth cable screw and the wiring harness clip. Make sure that the bolt fitted at top centre and the two bolts which hold the mounting bracket to the adaptor are removed.

20   Working at the lower part of the housing, remove the nut from the recess and the nut pointing in the reverse direction to the others.

21   Lower the engine/transmission until the starter motor mounting flange is below the engine mounting bracket on the front crossmember.

22   Separate the flywheel housing from the adaptor plate.

23   Withdraw the gearbox clutch shaft bearing inner race.

24   Push the power unit towards the left-hand valance and withdraw the flywheel housing from below the car. Remove the housing gasket.

25   Refitting is a reversal of removal, but tighten all nuts and bolts to the specified torque. Use a new housing flange gasket (photo).

26   If a new flywheel housing has been fitted then the idler gear endfloat must be checked after removing the idler bearing cap. To do this, pull the idler gear fully out and insert a rod into the centre of the gear. If a dial gauge is now fixed to the bearing cap (temporarily held

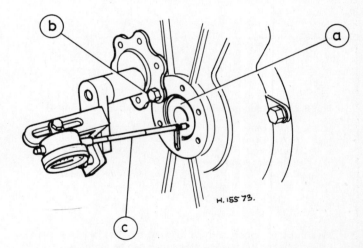

Fig. 6.5 Checking idler gear endfloat (Sec 6)

a   Idler gear   b   Cap fixing bolt   c   Rod and dial gauge

by one bolt) and the idler gear pushed fully inwards, the endfloat can be measured.

27   If the endfloat is not within a tolerance of 0.002 to 0.005 in (0.05 to 0.13 mm) then change the thrust washer for one of the four alternative thicknesses available. These are:

0.128 to 0.129 in (3.25 to 3.28 mm)
0.130 to 0.131 in (3.30 to 3.33 mm)
0.132 to 0.133 in (3.35 to 3.38 mm)
0.134 to 0.135 in (3.40 to 3.43 mm)

28   Use a new gasket when fitting the bearing cap.

29   Remove the filler plug from the top of the primary drive cover and pour in 1.5 pints (0.8 litre) of engine oil (photos).

30   Fill the engine/transmission with oil up to the full mark on the dipstick.

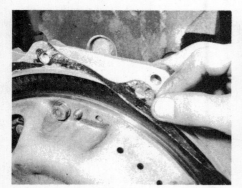

6.25 Flywheel housing gasket

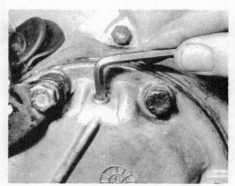

6.29A Primary drive cover filler plug

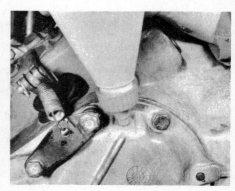

6.29B Priming primary drive with oil

8 Gearbox – dismantling (general)

1   Purchase new gaskets, lockwashers and circlips in advance of dismantling to save time. Observe strict cleanliness in all operations.
2   Unless suitable extractors or a press are available it is not recommended that some of the major assemblies are dismantled but the work should be left to your dealer.

7 Gearbox – removal and refitting

1   Removal of the gearbox is carried out in conjunction with the engine as a combined unit. The gearbox is separated from the engine after removal.
2   The procedures are fully described in Chapter 1.

9 Gearbox – dismantling into major assemblies

1   With the engine/gearbox removed from the car and the gearbox separated from the engine, detach the final drive, as described in the following paragraphs.
2   Remove the setscrews which secure the end covers to the final drive housing and gearcasing and pull off the covers.
3   Extract the differential bearing pre-load shims which are fitted beneath the end cover at the clutch end of the final drive housing.
4   Unscrew the single nut which holds the gear selector housing and the final drive housing to the gearcase. This nut is the one on the centre stud of the three which are located at the lower edge of the selector housing cover.
5   Unscrew and remove the remaining final drive housing to gearcase nuts. Lift off the exhaust mounting bracket and the final drive

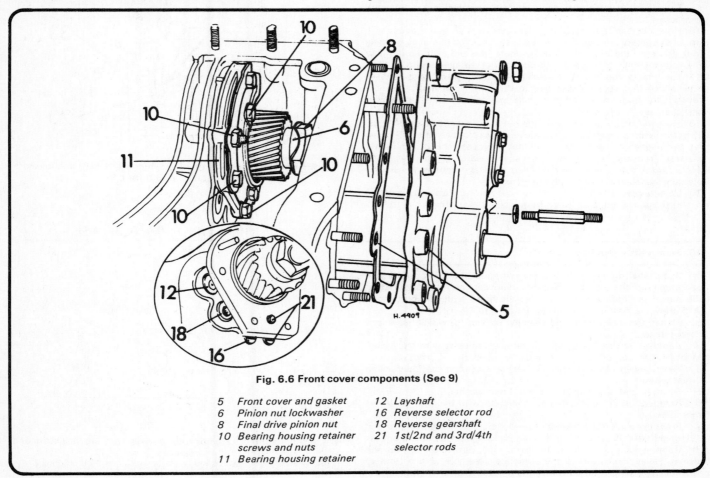

Fig. 6.6 Front cover components (Sec 9)

5 Front cover and gasket	12 Layshaft
6 Pinion nut lockwasher	16 Reverse selector rod
8 Final drive pinion nut	18 Reverse gearshaft
10 Bearing housing retainer	21 1st/2nd and 3rd/4th
screws and nuts	selector rods
11 Bearing housing retainer	

housing. Lift out the differential assembly with crownwheel.

6　Remove the gearbox front cover and gasket after unscrewing the securing nuts. Note the pick-up tube and filter screen assembly attached to the inside of the front cover which supplies oil to the crankshaft mounted oil pump. The pick-up housing can be dismantled by unscrewing the securing nut and then pulling the components apart. Clean the internal magnets and filter mesh of swarf and dirt. When reassembling, note the location of the three fibre washers. One fits at the base of the recess, the second one goes between the magnets and the oil pick-up body and the third one under the securing nut. The speedometer drive which engages with the end of the mainshaft may also be dismantled from the front cover if worn gears or components must be renewed.

7　If the mainshaft is to be dismantled, flatten the lockwasher tab on the final drive pinion nut and move the synchro sleeves to engage 1st and 3rd gears at the same time as a means of locking up the mainshaft, then slacken the final drive pinion nut. *This nut has a left-hand thread* and is very tight so use a socket with a long knuckle bar and have an assistant to hold the gearcase down, having first turned it upside down to make it more stable while the nut is being released.

8　Move the synchro sleeves back to disengage the gears.

9　Unlock and remove the bearing housing retainer screws and withdraw the retainer from the front cover end of the gearbox. Seven screws and one nut are involved in this operation.

10　Drive out the layshaft so that it emerges from the primary drive end of the gearcase. Lift out the laygear and thrust washers. Extract the bearing cages.

11　Remove the circlip and distance piece which retains the input shaft bearing.

12　Extract the selector rod lockplate.

13　Drive out the reverse selector rod so that it emerges from the primary drive end of the gearcase. During this operation, cover the gearcase open areas to prevent the interlock balls from being lost when they are ejected. The selector rod has two grooves in it.

14　Push the reverse gear idler shaft from the gearcase.

15　Extract the reverse idler gear and reverse selector fork from the gearcase.

16　Drive out 1st/2nd and 3rd/4th selector rods so that they emerge from the primary drive end of the gearcase. Each rod has three grooves in it, those on the 3rd/4th rod being nearer the end of the rod. Allow the selector forks to remain in the bottom of the gearcasing.

17　To remove the mainshaft assembly, tap a piece of hardwood against the 1st speed gearwheel to drive the mainshaft assembly towards the front cover. The mainshaft bearing housing will then clear the gearcase.

18　Take the weight of the gear train and extract the selector forks. Turn the bearing housing to clear the gearcase front aperture and then lift the mainshaft assembly out of the gearcase. The geartrain will separate from the input shaft at 4th speed gear.

19　Retrieve the selector fork locating balls and springs (previously ejected) from the bottom of the gearcase.

20　Using a plastic faced hammer, tap out the input shaft by applying a drift to its bearing outer track from inside the gearcase.

10 Mainshaft – overhaul

1　If the final drive pinion nut has not already been released (see paragraph 7 of the preceding Section) remove it now, making sure that the pinion is gripped in a vice which is fitted with jaw protectors. Withdraw the final drive pinion (Refer now to Fig. 6.10).

2　Flatten the lockwasher and unscrew the nut (6) which has a left-hand thread.

3　Remove the 3rd/4th synchro assembly complete with baulk ring. Mark its direction of fitting with a dab of quick drying paint.

4　Withdraw the shaft sleeve, 3rd speed gear and the interlocking thrust washer.

5　Remove 2nd speed gear and the thrust washer.

6　Remove 1st/2nd synchro unit complete with baulk rings. Invert the shaft.

7　Press the shaft from the bearing and then remove reverse gear and 1st speed gear. If the bearing is worn press it from its housing.

8　If the synchro units are to be dismantled, push the hub from the sleeve, having first covered the complete assembly with a cloth to prevent the springs and balls being lost when they are ejected.

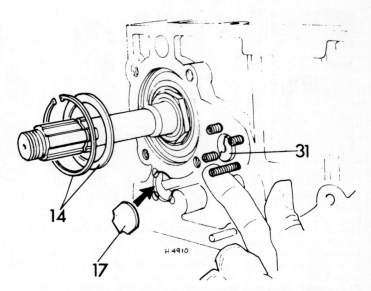

Fig. 6.7 Input shaft circlip and distance piece (14) selector rod locating plate (17) and layshaft (31) (Sec 9)

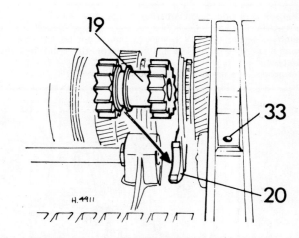

Fig. 6.8 Reverse gear (19) reverse selector fork (20) and layshaft lubrication hole (33) (Sec 9)

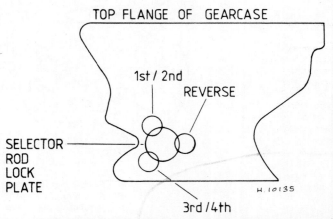

Fig. 6.9 Location of selector rods viewed from the lock plate end (Sec 9)

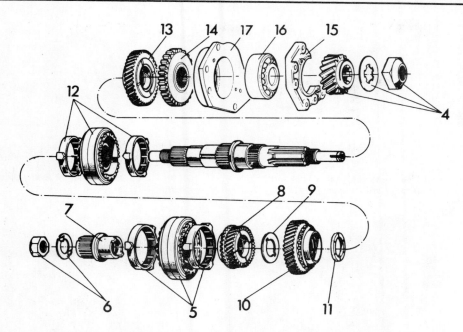

Fig. 6.10 Mainshaft components (Sec 10)

4   Final drive pinion, nut and lockplate
5   3rd/4th synchro unit
6   Shaft nut and lockplate
7   Shaft sleeve
8   3rd speed gear
9   Thrust washer
10 2nd speed gear
11 Thrust washer
12 1st/2nd synchro unit
13 1st speed gear
14 Reverse gear
15 Bearing retainer
16 Bearing
17 Bearing housing

9   Check all components for wear, especially the baulk rings. Push them onto the gear cones and check that they engage the cone before contacting the edge of the gear itself. If they do not, renew the baulk rings (photo).

10 When reassembling the synchro hub to the sleeve, use a worm drive clip or piston ring compressor to keep the springs and balls compressed. Align the hub to the sleeve and have the clip or compressor resting on the sleeve, then press the hub quickly out of the clip into the sleeve so that the balls and springs will not have time to be ejected. If necessary, a small screwdriver can be pushed down each of the hub grooves to depress the balls even further, but keep the clip or compressor in position as well.

11 If the bearing has been renewed, press it fully into its housing until it seats.

12 Refit 1st speed gear and reverse gear to the mainshaft (photos).

13 Press the bearing housing onto the shaft. Invert the shaft (photo).

14 Refit 1st/2nd synchro unit complete with baulk rings so that the two grooves on the sleeve are towards 2nd gear. This synchro unit is slightly smaller in diameter than the 3rd/4th synchro unit (photo).

15 Refit the thrust washer and 2nd speed gear (photos).

10.9 Mainshaft stripped

10.12A Fitting 1st speed gear to mainshaft

10.12B Fitting reverse gear to mainshaft

10.13 Fitting mainshaft bearing

10.14 Fitting 1st/2nd synchro

10.15A Fitting 2nd speed gear thrust washer

10.15B Fitting 2nd speed gear

10.16A 3rd speed gear, sleeve and interlocking thrust washer

10.16B Fitting 3rd speed gear assembly to mainshaft

10.16C 3rd speed gear assembly fitted

10.17 Fitting 3rd/4th synchro

10.18A Lock washer

10.18B Screwing on mainshaft nut

10.18C Nut secured with lock tab

10.19A Fitting final drive pinion to mainshaft

10.19B Pinion nut lockplate

10.19C Pinion nut fitted

16 Insert the sleeve into 3rd speed gear and position the interlocking thrust washer onto the sleeve. Refit the assembly to the mainshaft (photos).

17 Refit 3rd/4th synchro unit complete with baulk ring. Make sure that its fitted direction on the shaft is as recorded at dismantling. Once fitted, do not move the synchro sleeves or they will fall apart (photo).

18 Apply thread locking fluid to the mainshaft threads, use a new lockwasher and tighten and lock the nut, remembering that it has a left-hand thread. Use jaw protectors when gripping the shaft in the vice. Bend up the locktab (photos).

19 Fit the final drive pinion, lockwasher and nut. Tighten this nut (left-hand thread) after the mainshaft has been fitted into the gearbox (photos).

11 Input shaft – overhaul

1 Extract the mainshaft spigot bearing from inside the input shaft.

2 Grip the shaft in a vice fitted with jaw protectors and after flattening the lockwasher, remove the bearing retaining nut.

3 Press the shaft from the bearing.

4 Inspect all components for wear and renew as necessary.

5 To reassemble, press the input shaft into the bearing, refit the lockwasher and bearing retaining nut and tighten to the specified torque. Lock the nut (photos).

6 Fit the spigot bearing into the shaft (photo).

12 Gearbox – reassembly

1 Insert the detent springs and balls into their holes in the selector forks. Hold them in position by depressing them with a thin screwdriver and then insert a short, not more than $\frac{1}{2}$ in (12.7 mm) in length, dummy rod to slide over the detent ball and spring and retain them. If these rods are any longer, they cannot be ejected when the selector rods are pushed through the forks owing to lack of space

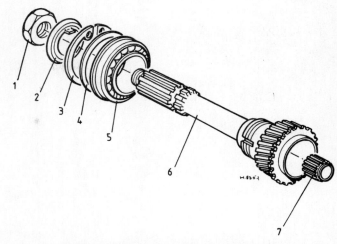

Fig. 6.11 Input shaft components (Sec 11)

1 Nut
2 Tab washer
3 Circlip
4 Distance piece
5 Ball bearing
6 Input shaft
7 Needle roller bearing

between the end of the fork and the centre web of the gearcase (photos).

2 Place the 1st/2nd and 3rd/4th selector forks in the bottom of the gearcase in their correct relative positions.

3 Locate the input shaft loosely in position, then pass the mainshaft assembly through the aperture in the front of the gearcase, turning the bearing housing as necessary and taking care not to foul the selector forks (photos).

11.5A Input shaft

11.5B Input shaft nut

11.6 Input shaft spigot bearing

12.1A Selector fork detent spring

12.1B Selector fork detent ball

12.3A Installing input shaft

12.3B Installing mainshaft assembly

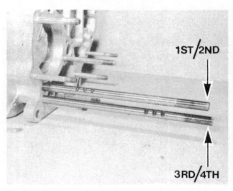

12.5 Selector rods

12.6A Fitting reverse idler shaft

12.6B Reverse idler gear and shaft fitted

12.8 Selector rod locating plate

12.9A Input shaft bearing distance piece

12.9B Input shaft bearing circlip being fitted

12.9C Input shaft bearing circlip fitted

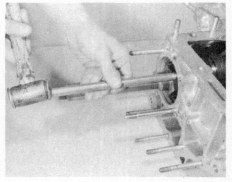

12.10 Tapping input shaft fully home

4   Locate the baulk ring on the face of 3rd/4th synchro unit.
5   Fit the 1st/2nd and 3rd/4th selector rods from the primary drive end of the gearcase, picking up their selector forks in the process and displacing the temporary detent ball rods. Retrieve these rods once the selector rods are fully in position with their end cut-outs correctly set to mate with the selector rod locating plate (photo).
6   Hold the reverse idler gear and shift fork in their correctly orientated position and insert the reverse idler shaft so that its cut-away end engages with the mainshaft bearing retainer plate (photos).
7   Insert the reverse selector rod, displacing the temporary detent ball retaining rod as it is pushed through the shift fork. Retrieve the temporary rod.
8   Fit the selector rod locating plate (photo).
9   The circlip and distance piece should now be fitted to the input shaft bearing. If a new bearing has been fitted, a distance piece must be selected from the sizes available until the circlip will just not enter

its groove. Now select and fit a distance piece of the next size down. Five selective size distance pieces are available (photo):

0.117 to 0.118 in	2.97 to 3.00 mm
0.121 to 0.122 in	3.07 to 3.10 mm
0.125 to 0.126 in	3.17 to 3.20 mm
0.129 to 0.130 in	3.28 to 3.30 mm
0.133 to 0.134 in	3.38 to 3.40 mm

10 Tap the input shaft assembly into position in the gearcase, applying pressure to the bearing outer track (photo).
11 If the laygear has been renewed, always fit a new split sleeve and circlip, also new thrust springs. Fit the needle roller bearing cages into each end of the laygear (photos).
12 Insert the layshaft into the casing so that the shaft cut-away is towards the clutch end and so positioned that it will locate in the adaptor plate (photos).

12.11A Laygear circlip

12.11B Laygear needle roller bearing

12.12A Layshaft cut-away

12.12B Adaptor plate cut-out for layshaft

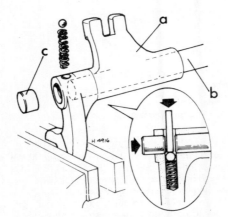

Fig. 6.12 Selector fork components (Sec 12)

a   *Fork*   c   *Temporary dowel*
b   *Selector rod*

13 Push the layshaft in so that it picks up the large thrust washer and the laygear and then select a small thrust washer of suitable thickness to provide a laygear endfloat of between 0.002 and 0.003 in (0.05 and 0.08 mm). Feeler blades can be used to check this. Thrust washers are available in selective thicknesses as follows (photos):

0.119 to 0.121 in	*3.02 to 3.07 mm*
0.123 to 0.125 in	*3.12 to 3.17 mm*
0.126 to 0.128 in	*3.21 to 3.26 mm*
0.130 to 0.132 in	*3.30 to 3.35 mm*
0.133 to 0.135 in	*3.38 to 3.43 mm*

14 Drive the layshaft fully home. Oil the geartrain and layshaft (photo).
15 Fit the bearing housing retainer to the front cover end of the gearcase. Use a new lockplate and bend up the tabs (photo).
16 Move 1st/2nd and 3rd/4th synchro sleeves to lock up the gears and to prevent the mainshaft turning.
17 Screw up the final drive pinion nut and lockwasher, having applied thread locking compound to the shaft threads. Tighten to the specified torque remembering that **the nut has a left-hand thread**. This nut has a high torque wrench setting. In the absence of a suitable wrench, use a socket or box spanner with a lever 18 in (457 mm) in length (photo).

12.13A Layshaft large thrust washer

12.13B Layshaft small thrust washer

12.13C Checking laygear endfloat

12.14 Geartrains installed

12.15 Mainshaft bearing housing retainer

12.17 Bearing housing retainer bolted in position

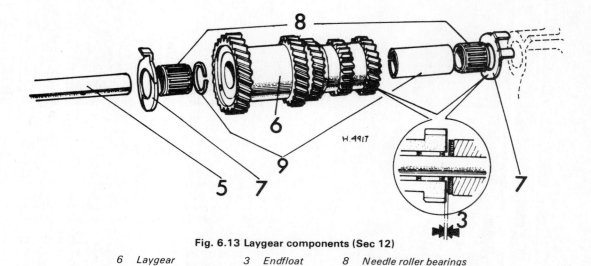

Fig. 6.13 Laygear components (Sec 12)

6 Laygear	3 Endfloat	8 Needle roller bearings
7 Thrust washers	5 Layshaft	9 Circlip and split sleeve

18 Bend up the lockplate to secure the nut.
19 Fit the gearbox front cover complete with oil pick-up assembly. Always use a new flange gasket and make sure that the speedo drive tongue engages correctly (photos).
20 Refit the differential and crownwheel assembly. Place a new gasket on the casing flange (photos).
21 Refit the final drive housing. When doing this, it is very important that the synchro sleeves are again in neutral and that the dogs on the forks are in alignment (photos).
22 Fit the differential bearing preload shims at the clutch end of the final drive housing and then bolt on the end covers, using new gaskets.

Fit the exhaust bracket to the (clutch) end cover. Make sure that the cover and gasket oilways are aligned (photos).
23 Fit the selector lever/cover assembly. It is again very important to have the dogs and levers in perfect alignment during assembly. A worn oil seal will allow the levers to fall out of alignment during fitting and a new seal should be fitted beforehand (photos).
24 Oil all gears and friction surfaces before the gearbox is connected to the engine and check that all nuts and bolts have been tightened to the specified torque.
25 The transmission is now ready for connecting to the engine as described in Chapter 1, Section 24.

12.19A Fitting front cover

12.19B Speedo drive tongue engagement

12.19C Front cover in position

12.20A Final drive/differential

12.20B Final drive casing gasket

12.21A Selector dogs in alignment

12.21B Fitting final drive housing

12.22A Differential bearing end cover gasket

12.22B Differential bearing preload shim

12.23A Selector shaft oil seal

12.23B Selector dogs in alignment

12.23C Selector cover in position

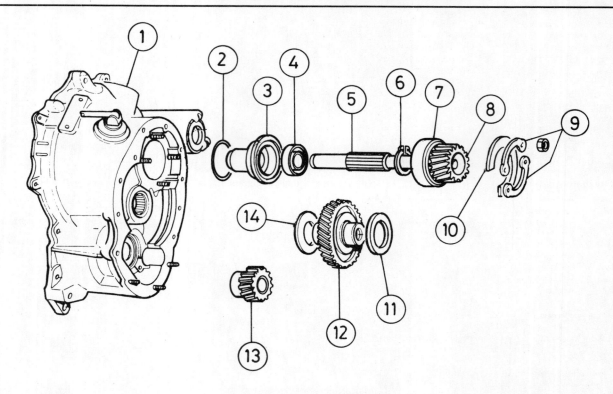

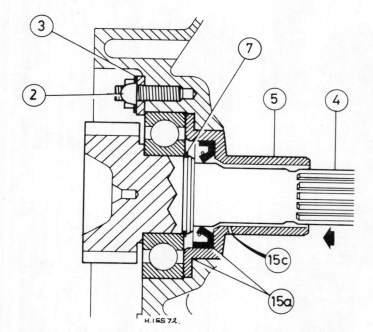

Fig. 6.14 Primary drive geartrain (Sec 13)

1	Flywheel housing	5 Clutch shaft	9 Lockplate	12 Idler gear
2	O-ring	6 Circlip	10 Retaining plate	13 Input shaft gear
3	Oil seal retainer	7 Bearing	11 Thrust washer	14 Washer
4	Oil seal	8 Clutch shaft gear		

Fig. 6.15 Sectional view of clutch shaft components (Sec 13)

2 Lockplate and nut
3 Retaining plate
4 Clutch shaft
5 Oil seal retainer
7 Bearing circlip
15a Oil seal
15c Oil seal retainer drain hole

13 Primary drive geartrain — overhaul

1   The following operations may be carried out without removing the gearbox from the car.
2   Remove the flywheel housing as described in Section 6.
3   Remove the retaining nuts and lockwashers and take off the clutch shaft bearing retainer.
4   Press the clutch shaft from the housing.
5   Extract the oil seal retainer.
6   Remove the O-ring and the oil seal.
7   Extract the bearing circlip.
8   Support the bearing and press the clutch shaft from it.
9   Remove the input shaft bearing and outer track.
10 Extract the idler gear bearings.
11 From inside the flywheel housing remove the clutch release components as described in Chapter 5.
12 Reassembly is a reversal of dismantling, but grease the lips of the new oil seals and make sure that the drain hole in the oil seal retainer is aligned with the one in the flywheel housing.
13 Refer to Section 6 for details of refitting the flywheel housing and refilling with oil.

PART 2: AUTOMATIC TRANSMISSION

15 General description

The automatic transmission is of three speed type mounted transversely at the base of the engine crankcase. The final drive/differential is attached to the main transmission casing.
   The torque converter is not mounted in line with the geartrain but fitted above it. The torque developed by the engine is transmitted to

the geartrain by a special type of chain. With this layout there is a division between the engine and the automatic transmission unit so that the engine oil is retained in the engine and normal automatic transmission fluid in the automatic transmission and differential unit. This is of course a deviation from normal Austin-Rover transverse unit design.

Owing to the complexity of the automatic transmission unit, if performance is not up to standard or overhaul is necessary, it is imperative that this be left to your local main agents who will have all the special equipment and knowledge for fault diagnosis and rectification. The successful overhaul of an automatic transmission unit requires the use of many very special tools and the content of this Chapter is therefore confined to supplying general information and any service information that will be of practical use to the owner.

16 Fluid level checking

1 The most important item of regular maintenance is to keep the fluid level topped up to the specified level.
2 Start the engine and run it until it reaches normal operating temperature.
3 Move the speed selector lever to each selector position, pausing briefly at each position and finally locate it in Park with the engine idling.
4 Withdraw the dipstick, wipe it clean, re-insert it fully and withdraw it again.
5 Read off the fluid level which should be between the MIN mark and the notch on the WARM side.
6 The fluid level may also be checked after the car has just come in from a run of between 15 and 20 miles (25 and 30 km).
7 Withdraw the dipstick with the engine still idling, wipe it clean, re-insert it and withdraw it again. The fluid level should be between the notch and the MAX mark on the HOT side.
8 Where required, top up with the specified fluid by pouring it into the dipstick guide/filler tube until it reaches the notch if the level was checked on the WARM side or the MAX mark if checked on the HOT side.
9 Although draining the automatic transmission and refilling with fresh fluid is not a service item specified by the manufacturers, it is recommended that this is carried out at the intervals specified in Routine Maintenance in the interest of smooth operation and long life of the unit.
10 Drain the fluid hot by unscrewing the drain plug provided, but avoid contact with the fluid as it will be very hot.
11 Pour in only the specified quantity of fresh fluid of recommended type.

17 Speed selector cable – adjustment

1 Should the automatic gearbox fail to respond to a selector position, first check the fluid level and then carry out the following test.
2 With the engine idling in N, select D and release the handbrake. Drive forward by accelerating and then push the selector lever to N. Disconnection of the drive should be felt immediately.
3 Repeat the operation in reverse using R and N.
4 Any malfunction may be due to faulty selector cable adjustment which should be rectified as follows. Switch off the engine, position the selector lever in 1. Slacken the outer cable adjuster locknut. Check that the outer cable has free movement and can be pushed and pulled in and out. Grip the outer cable and pull it from the selector housing until the valve rod is felt to contact the stop in the valve block. Hold the cable in this position and tighten the locknut.
5 Check that the engine will only start in the N or P positions.

18 Speed selector cable – renewal

1 To renew the cable, select P and then release the selector cable retaining clips. Use a thin screwdriver to lever against the inner clip and pull the cable free.
2 Unscrew the outer cable nut and withdraw the cable as far as possible. Unscrew the inner cable to disengage it from the selector lever fork.
3 Detach the cable from the support clip on the body. Fitting the

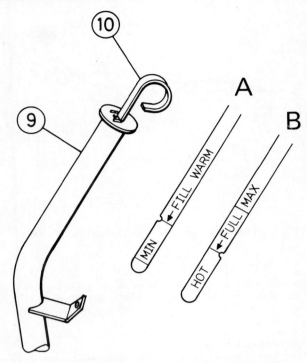

Fig. 6.16 Automatic transmission dipstick (Sec 16)

A Warm side   *9 Guide/filler tube*
B Hot side   *10 Dipstick*

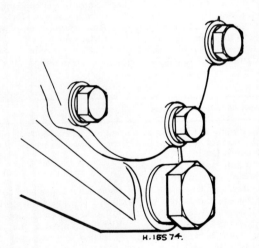

Fig. 6.17 Transmission drain plug (Sec 16)

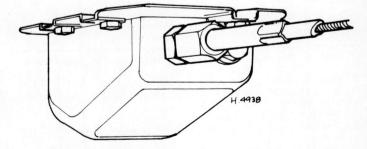

Fig. 6.18 Speed selector cable and locknut (Sec 17)

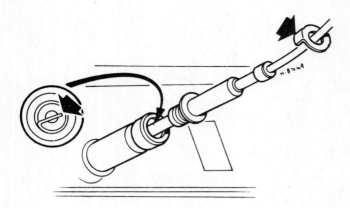

Fig. 6.19 Speed selector cable retaining clips (Sec 18)

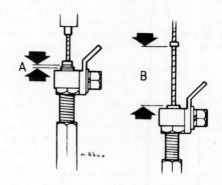

Fig. 6.20 Kickdown cable adjustment at carburettor (Sec 19)

A $\frac{1}{8}$ in (3.0 mm) at idle   B   1.75 in (44.5 mm) at full throttle

new cable is a reversal of removal but make sure that the inner cable wire clip springs into position before refitting the outer retaining clip. Adjust the selector cable as described in the preceding Section.

19 Kickdown cable – adjustment

1   Select P, then start the engine and allow it to idle at 750 rev/min.
2   Check that the crimped stop on the downshift cable is positioned to give a clearance A (Fig. 6.20) between the stop and the end face of the threaded adjuster of approximately $\frac{1}{8}$ in (3.0 mm).
3   Stop the engine, then depress the accelerator pedal fully and have an assistant check that the carburettor throttle lever is fully open against its stop. Check that the dimension B between the stop and the end face of the threaded adjuster is $1\frac{3}{4}$ in (44.5 mm).
4   Adjust as necessary by releasing the locknut on the threaded adjuster and turning the adjuster as required. Re-tighten the locknut on completion.

20 Front and rear brake bands – adjustment

1   This adjustment should only be carried out when any fault may be the result of incorrect setting of one or both brake bands as detailed in Fault Diagnosis (Section 26).
2   Drain the automatic transmission.
3   Remove the servo cover and gasket.
4   Release the now accessible front band adjusting screw locknut and tighten the adjusting screw to 8 lbf ft (11 Nm). Now unscrew the adjusting screw by between 1 and $1\frac{1}{4}$ turns. Retighten the locknut.
5   Where the rear band requires adjustment, unhook the piston return spring from the servo casing and release the adjusting screw locknut. Tighten the adjusting screw to 10 lbf ft (1.2 Nm) only. Unscrew the adjusting screw by between 2 and $2\frac{1}{2}$ turns. Retighten the locknut. Refit the piston return spring.
6   Refit the servo cover and a new gasket and refill the automatic transmission to the correct level.

21 Speed selector lever – removal and refitting

1   Remove the console from inside the car as described in Chapter 12.
2   Release the speed selector knob and locknut.
3   Prise the escutcheon plate moulding from its location and disconnect the indicator lamp wires.
4   Remove the front carpet.
5   Release the automatic selector cable retaining clip and pull the cable free.
6   Extract the six screws which hold the speed selector mechanism to the floor and withdraw the mechanism from under the floor of the car.

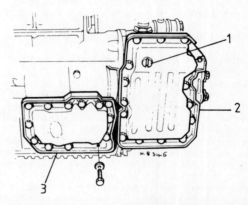

Fig. 6.21 Automatic transmission bottom covers (Sec 20)

1   Fluid drain plug   3   Servo cover
2   Valve body cover

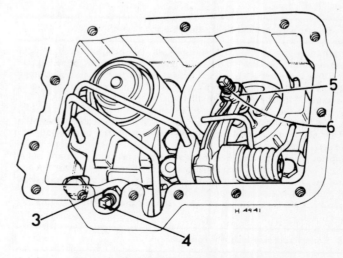

Fig. 6.22 Band adjusting screws (Sec 20)

3   Locknut   5   Locknut
4   Front band adjuster screw 6   Rear band adjuster screw

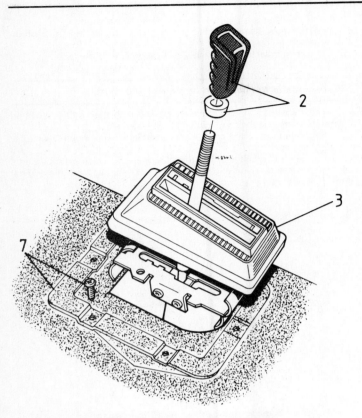

Fig. 6.23 Speed selector lever (Sec 21)

2   Knob and locknut   7   Self-tapping screw
3   Escutcheon plate moulding

7   Refitting is a reversal of removal. Do not kink the cable or stiff gearchanging will result when it is refitted. Adjust as described in Section 17.

22 Starter inhibitor/reverse lamp switch – removal and refitting

1   Working inside the car, peel back the carpet under the heater and disconnect the two inhibitor switch leads.
2   Press the grommet out of the body panel and push the leads through.
3   Raise the front right-hand side of the car and release the switch wiring harness from the automatic transmission fluid filler tube and suspension pipe.
4   Disconnect the switch leads, remove the switch fixing bolt and withdraw the switch.
5   Refitting is a reversal of removal.

23 Primary drive chain – removal and refitting

1   The primary drive chain used in the automatic transmission is accessible without having to remove the engine/transmission from the car, but the following preliminary work must first be carried out.
2   Drain the transmission fluid.
3   Remove the battery.
4   Unbolt and remove the primary chain cover and its gasket.
5   Flatten the tabs on the chain sprocket lockwashers.
6   Unscrew the nut from the torque converter sprocket. To prevent the sprocket from turning, insert a wedge between the sprocket and the front pump suction boss.
7   Unscrew the input sprocket nut.

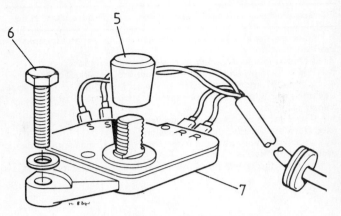

Fig. 6.24 Starter inhibitor switch (Sec 22)

5   Protective cap   7   Switch body
6   Setscrew

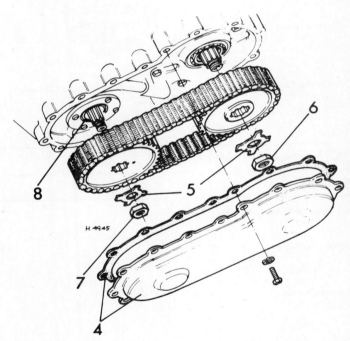

Fig. 6.25 Primary drive components (Sec 23)

4   Chain cover and gasket   7   Input shaft nut
5   Lockplates   8   Input shaft spacer and
6   Torque converter shaft nut   bearing

8   Pull the sprockets and chain from the shafts.
9   Refitting is a reversal of removal, but due to the restricted space available, when attempting to fit a torque wrench to the torque converter sprocket nut, the following work should first be carried out.
10   Support the right-hand end of the gearbox on a jack.
11   Unscrew the through bolt from the right-hand front mounting of the engine.
12   Remove the engine right-hand rear mounting support bracket.
13   Raise the engine/transmission to give enough clearance for the torque wrench.

24 Speedometer drivegear and pinion – removal and refitting

1   Unscrew the knurled ring and detach the speedometer cable from the transmission.
2   Remove the spring plate, screwed bush and gasket.
3   Pull the pinion and pinion bush from the housing.
4   If the drivegear must be renewed, first drain the automatic transmission fluid, and disconnect the speedometer cable from the transmission.
5   If the car is equipped with power steering, release the support clips from the stud on the transmission.
6   Unbolt and withdraw the extension housing, release the governor pressure tube from its seal.
7   Remove the speedometer drivegear. The best way to do this is to use a nut splitter with care to avoid marking the shaft.
8   Fit the new gear (shoulder leading) by tapping it into position using a soft-faced mallet.
9   The remainder of reassembly is a reversal of removal.

25 Automatic transmission – removal and refitting

1   The transmission is removed together with the engine and then separated as described in Chapter 1, Section 14.
2   Reconnection to the engine and refitting are described in Chapter 1, Section 25.
3   If a new or rebuilt unit is being fitted, check which external items come with the new unit and remove any not supplied from the original unit before parting with it.

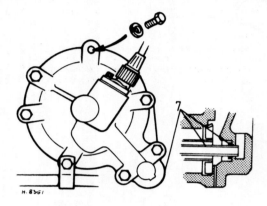

Fig. 6.26 Extension housing showing governor pressure tube and seal (7) (Sec 24)

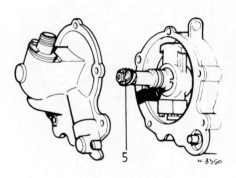

Fig. 6.27 Speedometer drivegear (5) on transmission shaft (Sec 24)

14 Fault diagnosis – manual gearbox

Symptom	Reason(s)
Weak or ineffective synchromesh	Synchronising cones worn, split or damaged Baulk ring synchromesh dogs worn or damaged
Jumps out of gear	Broken gearchange fork rod spring Gearbox coupling dogs badly worn Selector fork rod groove badly worn
Excessive noise	Incorrect grade of oil in gearbox or oil level too low Bush or needle roller bearings worn or damaged Gear teeth excessively worn or damaged Layshaft thrust washers worn allowing excessive endplay
Excessive difficulty in engaging gear	Clutch fault (see Chapter 5)

26 Fault diagnosis – automatic transmission

Symptom	Reason(s)
Engine will not start in N or P	Flat battery Fault in circuit Incorrect selector cable adjustment Faulty inhibitor switch
Engine starts in positions other than N or P	Incorrect selector cable adjustment Faulty inhibitor switch
Severe bump when selecting D or R	Idling speed too high Faulty downshift valve or cable adjustment

Symptom	Reason(s)
Poor acceleration and low maximum speed	Incorrect fluid level Incorrect selector cable adjustment
Delayed or no 1 to 2 shift	Incorrect front brake band adjustment
Delayed or no 2 to 3 shift	Incorrect front brake band adjustment
No 3 to 2 downshift or engine braking	Incorrect front brake band adjustment
Drag in R	Incorrect front brake band adjustment
Drag in D or 2	Incorrect rear brake band adjustment
No 2 or 1 downshift or engine braking	Incorrect rear brake band adjustment
Slip on take off in R and no engine braking in 1	Incorrect rear brake band adjustment

The most likely causes of faulty operation are incorrect oil level and linkage adjustment. Any other faults or mal-operation of the automatic transmission unit must be due to internal faults and should be rectified by your dealer. An indication of a major internal fault may be gained from the colour of the oil which under normal conditions should be transparent red. If it becomes discoloured or black then burned clutch or brake bands must be suspected.

Chapter 7 Final drive and differential

Contents

Specifications

Type .... Crownwheel and differential assembly integral with gearbox

Ratios
Manual transmission .... 3.722 : 1
Automatic transmission .... 3.830 : 1

Differential bearing preload .... 0.003 to 0.005 in (0.08 to 0.13 mm)

Torque wrench settings

	lbf ft	Nm
Crownwheel bolts	60	82
Differential end cover setscrews	18	24
Differential housing cover nuts:		
$\frac{5}{16}$ in	18	24
$\frac{7}{16}$ in	40	54
Final drive pinion nut	160	218

1 General description

The final drive is integral with the gearbox or automatic transmission and consists of helical gears and the differential.

The ratio differs between the units used on manual or automatic transmission (see Specifications).

On cars with a manual gearbox, lubrication of the final drive components is provided by the common engine/gearbox oil.

On cars with automatic transmission, the separate transmission fluid is used to lubricate the final drive.

2 Differential end cover oil seals – renewal

1 This work can be carried out without having to remove the final drive from the car. Raise the front of the car and support it securely.
2 Refer to Chapter 11 and disconnect the upper and lower suspension swivel joints.
3 Using the tool and method described in Chapter 8, release the driveshafts from the differential.
4 Reconnect the upper suspension swivel joint temporarily and support the inner end of the driveshaft on an axle stand.

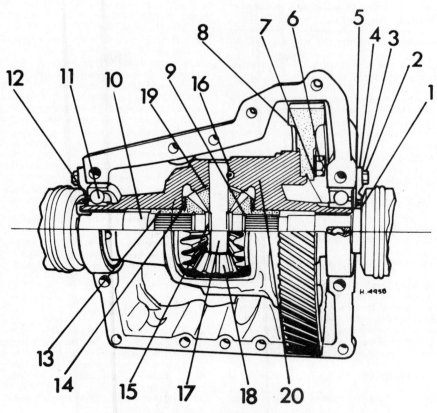

Fig. 7.1 Cutaway view of the final drive (Sec 1)

1	End cover oil seal	6	Crownwheel bolt	11	Differential case bearing	16	Tension pin
2	End cover setscrew	7	Lockwasher	12	Pre-load shims	17	Pinion centre pin
3	Lockwasher	8	Crownwheel	13	Differential gear	18	Differential pinion
4	End cover	9	Driveshaft retaining circlip	14	Washer	19	Washer
5	Gasket	10	Driveshaft	15	Distance piece	20	Differential case

5   Clean away any dirt or oil from the differential end covers.

6   If the oil seal on the end cover at the clutch (or torque converter end) is being renewed, remove the cover securing bolts and withdraw the cover carefully retaining the shims fitted against the thrust face of the bearing.

7   If the oil seal is being renewed on the crownwheel side, remove the exhaust steady bracket from the final drive housing flange before unbolting the end cover.

8   Lever the old oil seal from the end cover and carefully drive in a new one.

9   Clean away all traces of old gasket and locate a new one making sure that the oil cut-outs in both the cover and gasket are in alignment.

10 Return the shims to their original location and refit the cover, again ensuring that the cut-outs in the cover are in alignment with the oil holes in the differential housing.

11 Tighten the cover bolts to the specified torque.

12 Refit the driveshaft to the differential having first released the suspension upper swivel joint.

13 Engage the driveshaft with its retaining ring in the differential. Check that it is positively locked in engagement with the ring and if necessary, fit a worm drive clip to the inner sliding joint so that a drift can be applied to the edge of the clip and the driveshaft tapped fully into position.

14 Reconnect the suspension upper and lower swivel joints. Tighten the nuts to the specified torque, refit the roadwheel and then lower the car to the ground.

15 If there was any loss of lubricant during these operations, check and top-up the engine oil (manual gearbox) or the transmission fluid (automatic transmission).

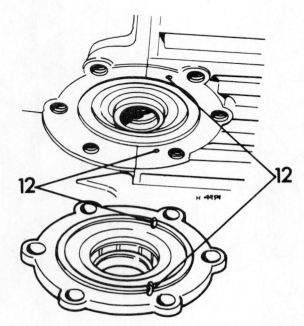

Fig. 7.2 Differential end covers showing oil cut-outs and holes (12) (Sec 2)

3 Final drive housing and differential (manual and automatic transmission) – removal, overhaul and refitting

1 Remove the engine/transmission from the car, as described in Chapter 1.

2 Remove the end covers from the final drive housing, carefully retaining the shims located under the cover at the clutch (or torque converter) end.

3 *On cars fitted with a manual gearbox,* remove the single nut which secures the selector lever cover and the final drive housing to the gearbox.

4 Unscrew and remove the remaining nuts which secure the final drive housing to the gearbox casing.

5 Remove the exhaust mounting bracket.

6 Pull off the final drive housing and lift out the differential assembly.

7 Mark the relative position of the crownwheel to the differential cage.

8 Flatten the tabs of the lockwashers and remove the bolts which secure the crownwheel to the cage.

9 Lift the crownwheel from the cage.

10 Pull the bearings from the differential cage using a suitable bearing extractor.

11 Drive out the tension pins which retain the differential pinion pin and then remove the pin.

12 Remove the pinions and thrust washers.

13 Remove the differential gears and washers.

14 Clean and examine all components for wear or damage and renew as necessary. If the crownwheel or the pinion is worn, both components must be renewed as a matched pair. Removal of the pinion is described in Chapter 6, Sections 9 and 10 for cars with manual gearbox and the next Section of this Chapter for cars having automatic transmission.

15 Commence reassembly by fitting new circlips into the retaining grooves in the differential gears.

16 Refit the differential gears and washers, the pinions and thrust washers using new tension pins.

17 Refit the bearings to the differential cage. Apply pressure to the bearing centre tracks only and make sure that the numbers on the bearings face away from the cage.

18 Bolt the crownwheel to the differential cage using new locking tabs. If the original crownwheel is being refitted, make sure that the marks made before dismantling are in alignment.

19 Refit the differential unit into the final drive housing.

20 *If the original differential cage bearings have been refitted,* bolt on the end covers, using new gaskets and making sure that the original shims are returned to their location under the end cover at the clutch (or torque converter) end of the final drive housing.

21 *If new differential cage bearings have been fitted,* only tighten the final drive housing securing nuts enough to just hold the unit to the gearbox so that it can be slightly displaced when the end cover is fitted at the clutch (or torque converter) end of the housing. Fit and bolt on the end cover and a new gasket at the crownwheel end. Fit the end cover at the clutch (torque converter) end without its gasket or any shims and tighten the bolts evenly just enough for the cover to nip the bearing outer track. Using feeler blades, measure the gap between the flange of the clutch (torque converter) end cover and the final drive housing and record the reading. Calculate the shim pack required by employing the formula (a-b) + c = d when -

(a) is average thickness of cover gasket (compressed) 0.008 in (0.20 mm)

(b) clearance measured (A) (Fig. 7.4) between cover and casing

(c) pre-load required 0.004 in (0.10 mm)

(d) thickness of bearing shim pack required

Shims are available in the following thicknesses:

0.0015 in (0.004 mm)
0.0025 in (0.006 mm)

22 To ensure accuracy in this bearing pre-load adjustment method, it is best to take measurements with feeler blades at several different points. If the gaps vary, this will indicate that the end cover screws at the clutch (torque converter) end have not been tightened evenly. Adjust them carefully until the same gap is recorded at all points of measurement.

23 Smear the selected shims with grease and fit them to the thrust face of the bearing.

24 Refit the end cover at the clutch (torque converter) end using a new gasket and making sure that the oil holes and channels in the cover, gasket and final drive housing are in alignment.

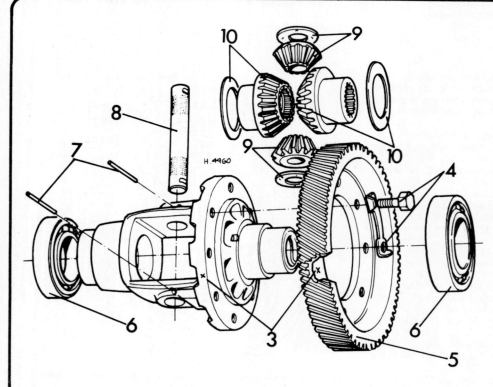

Fig. 7.3 Exploded view of final drive (Sec 3)

3   Crownwheel and case alignment marks
4   Crownwheel lockwashers and bolts
5   Crownwheel
6   Differential case bearings
7   Roll pins
8   Pinion pin
9   Pinions and thrust washers
10 Differential gears and washers

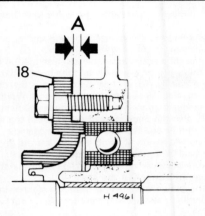

Fig. 7.4 Differential bearing preload calculation diagram (Sec 3)

A   Cover to flange clearance   *18   End cover*

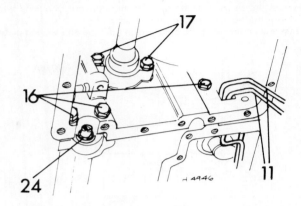

Fig. 7.5 Interior view of torque converter housing (Sec 4)

11   Rear clutch and servo tubes   *24   O-ring seal for governor*
16   Internal securing screws   *pressure tube*
17   End sealing plate screws

25 Tighten the end cover screws to the specified torque wrench settings and fit the exhaust mounting bracket.
26 The final drive should now be fitted to the gearbox casing by reversing the removal operations.
27 Refitting is a reversal of removal. Should the final drive unit not seat fully on the gearbox casing, this will probably be due to the crownwheel not meshing with the final drive pinion. To overcome this, just turn either gearwheel slightly. It is recommended that the selector lever/cover assembly is fitted separately after the final drive housing, with selector levers and shift fork dogs in perfect alignment.

4   Final drive pinion (automatic transmission) – removal and refitting

1 This operation will be required if the crownwheel in the final drive unit is to be renewed as the crownwheel and the final drive pinion must always be renewed at the same time as a matched pair.
2 Remove the engine/transmission from the car as described in Chapter 1.
3 Remove the distributor cap and leads.
4 Unbolt and remove the starter motor and retain the spacer.
5 Unscrew and remove the four screws which hold the driveplate to the torque converter. These are accessible one at a time through the starter motor aperture, the crankshaft having to be turned to bring them into view.
6 Remove the primary drive chain as described in Section 23 of Chapter 6.
7 Turn the engine/transmission onto its side and unbolt and remove the servo cover and the valve body cover.
8 Release the oil strainer retainer clip.
9 Remove the magnet from its location on the front servo tubes.
10 Release the rear clutch apply tube and let it hang.
11 Release the rear servo apply tube and let it hang.
12 Remove the common lubrication tube.
13 Remove the pick-up strainer, taking care not to pull the fluid pick-up pipe from its location in the front pump.
14 Extract the retaining clip for the front servo release tube and retrieve the sealing washer which fits between the clip and the casing.
15 Remove the front servo apply tube, then the front servo release tube.
16 Disconnect the throttle cable from the downshift cam.
17 Support the valve body with the hand and extract the three securing screws, noting the position of the oil strainer locating plate.
18 Withdraw the valve body. Disengage the control rod, but keep the oil pick-up tube connected to the front pump.
19 Extract the two bolts which hold the engine rear mounting bracket.
20 Extract the bolt which holds the wiring harness clip and the top of the converter housing to the adapter plate.
21 Extract the four bolts which hold the converter housing to the main transmission casing.

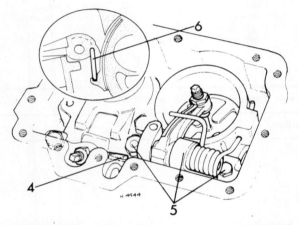

Fig. 7.6 Automatic transmission rear servo centre support dowel bolt (4) rear servo and securing screws (5), rear servo arm strut (6) (Sec 4)

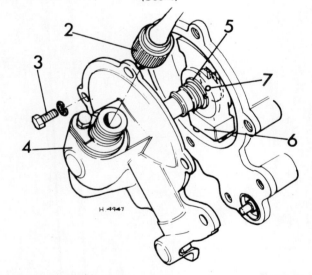

Fig. 7.7 Automatic transmission extension housing and governor (Sec 4)

2   Speedometer drive cable   *5   Governor circlip*
3   Setscrew   *6   Governor*
4   Extension housing   *7   Governor locating ball*

22 Extract the three bolts from inside the converter housing which also secure the housing to the main casing.
23 Unscrew and remove the bolts which hold the end sealing plate to

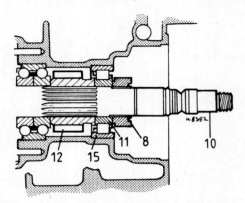

Fig. 7.8 Sectional view of automatic transmission output shaft and pinion (Sec 4)

8 Pinion nut	12 Final drive pinion
10 Output shaft	15 Pinion roller bearing
11 Pinion roller bearing	outer race
inner race	

the converter housing. Note that the outer bolt has a self-sealing washer.
24 Withdraw the torque converter housing and take off the spacer from the input shaft.
25 Remove the differential assembly as described in Section 3.
26 Remove the geartrain assembly and the front and rear brake bands from the main casing. To do this, first remove the dowel bolts and sealing washers which secure the centre support. Withdraw the centre support and planet carrier assembly. Remove the needle bearing and thrust washer, also the backing washer. Squeeze the ends of the rear brake band together and remove it from the casing.
27 Remove the extension housing and release the governor pressure tube from its seal in the extension housing.
28 Remove the governor retaining circlips from the shaft taking care not to damage the surface of the shaft.
29 Relieve the pinion nut staking from the depression in the shaft. Engage the parking pawl and unscrew the pinion nut which has a *left-hand thread.*
30 Drive the output shaft assembly out of its bearings, then remove the inner race of the pinion support roller bearing.
31 Using a suitable extractor, pull the drive pinion from the output shaft.
32 Refitting is a reversal of removal but observe the following points:

  (a) *Fit the inner race of the roller bearing after the input shaft has been refitted*
  (b) *Use a new pinion nut and tighten to the specified torque. Stake the nut*
  (c) *Retain the thrust washers in position using petroleum jelly*
  (d) *When refitting the governor, make sure that the cover plate is towards the pinion*

---

5   Fault diagnosis – final drive and differential

Symptom	Reason(s)
Oil leaks	Faulty end cover oil seals
	Faulty end cover gaskets
	End covers incorrectly fitted (oil holes not in alignment)
Noisy operation	General wear in components
	Incorrect bearing preload

Chapter 8 Driveshafts, hubs, roadwheels and tyres

Contents

Specifications

Driveshafts
Type .... Solid shaft with constant velocity (CV) joint at outboard end and sliding joint at inboard end. Shafts are of unequal length and not interchangeable

Driveshaft overall length (sliding joint compressed):
    Shorter shaft .... 30.75 in (781.1 mm)
    Longer shaft .... 31.50 in (800.1 mm)

Hubs
Bearing type:
    Front .... Two tapered roller bearings
    Rear .... Two ball bearings

Roadwheels
Type .... Pressed steel or alloy (Vanden Plas)
Size .... 4.5 J x 14

Tyres
Type .... Radial ply
Size .... 185/70SR x 14

Tyre pressures (cold)
Normal:
    Front .... 26 lbf/in^2 (1.8 bar)
    Rear .... 24 lbf/in^2 (1.7 bar)
When towing:
    Front .... 26 lbf/in^2 (1.8 bar)
    Rear .... 28 lbf/in^2 (1.9 bar)

Torque wrench settings

	lbf ft	Nm
Driveshaft nut	200	272
Front suspension upper swivel nut	40	54
Front suspension lower swivel nut	40	54
Tie-rod end balljoint nut	35	48
Rear hub nut	60	83
Roadwheel nuts	45	61

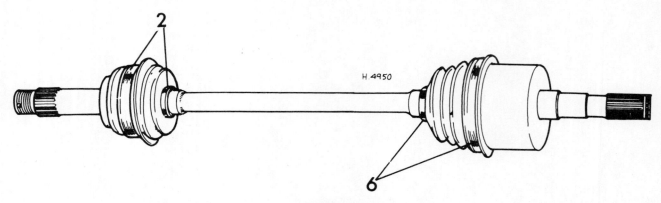

Fig. 8.1 Driveshaft (Sec 1)

2   Outboard (CV) joint   6   Inboard (sliding) joint

1   Description and maintenance

1   Drive is transmitted from the final drive differential unit to the front roadwheels by means of two driveshafts of unequal length.
2   The shafts are of solid construction incorporating constant velocity joints, the inboard joint being of sliding type while the outboard one is of ball and cage type.
3   No regular maintenance is required, but the flexible gaiters should be inspected at frequent intervals for splits or general deterioration and renewed where necessary as described in Section 3.

2   Driveshaft – removal and refitting

1   Insert a piece of wood approximately $1\frac{1}{2}$ in wide x $\frac{5}{8}$ in thick (40.0 x 16.0 mm) between the suspension upper arm and the rebound rubber, see Chapter 11.
2   Slacken the front roadwheel nuts.
3   Raise the front of the car and support it on a stand placed under each front side jacking point.
4   Remove the roadwheels.
5   Slacken the U-bolt which secures the exhaust pipe, remove the bracket from the differential housing and push the bracket down the pipe.
6   Release the driveshaft from the differential using a wide flat piece of metal as a lever. A special tool (No. 18G 1263) is available if your dealer can be persuaded to loan it to you.
7   Insert the tool at the junction of the driveshaft and differential (Fig. 8.2) and then strike the tool a sharp blow in the direction of the differential. **Do not** try to pull the driveshaft out of engagement with the differential as you will not release the retaining ring from the driveshaft groove unless a sharp blow is given.
8   Extract the split pin from the hub nut and remove the nut (Fig. 8.3).
9   In order to prevent the shaft turning, have an assistant apply the footbrake hard as the nut is unscrewed.
10 Using a soft-faced mallet, drive the shaft from the drive flange splines and then extract the split collar. Alternatively, use a three-legged puller with the centre screw applied to the end of the driveshaft.
11 Disconnect the tie-rod end from the steering arm (Chapter 11).
12 Disconnect the upper and lower suspension balljoints from the suspension arms and withdraw the driveshaft assembly. **Do not** pull the driveshaft by gripping the stub axle as it may cause the outer constant velocity joint to separate.
13 Extract the water shield and the bearing spacer and then press the shaft out of the inner bearing.
14 Commence refitting by fitting the inner bearing onto the driveshaft tight against the flange then fit the bearing spacer.
15 Position the water shield approximately $\frac{1}{4}$ in (6.0 mm) onto the

shaft and then pack the area between the shield and the bearing with grease.
16 Clean all the old grease from inside the hub and pack the area between the water shield and the oil seal with grease.
17 Push the driveshaft into the final drive (differential unit) with sufficient force to lock the shaft retaining ring. Check that it is properly locked by pulling it gently.
18 Reconnect the suspension and tie-rod end balljoints. Tighten the nuts to the specified torque.
19 Pull the driveshaft into the hub either using a suitable puller or by fitting pieces of tubing as distance pieces and screwing on the shaft nut.
20 Tighten all bolts and nuts to the specified torque and in the case of the shaft nut, turn it to the next hole (not backwards) if necessary to align the slot in the castellated nut. Use a new split pin.

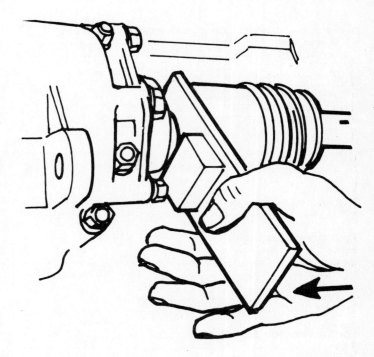

Fig. 8.2 Disengaging a driveshaft using the special tool (Sec 2)

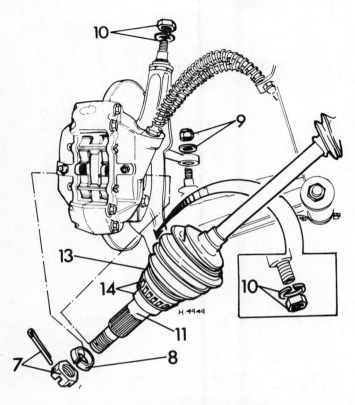

Fig. 8.3 Driveshaft attachments (Sec 2)

7   Nut and split-pin
8   Split collar
9   Tie-rod balljoint nut
10 Upper and lower suspension
    swivel joint nuts
11 Driveshaft
13 Water shield
14 Bearing and spacer

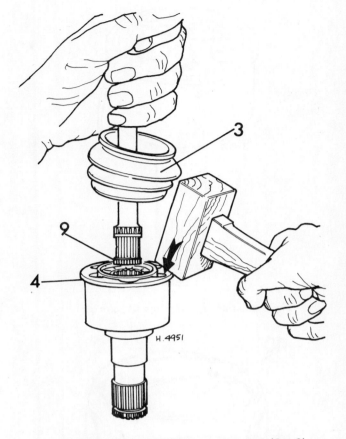

Fig. 8.4 Separating CV joint from driveshaft (Sec 2)

3   Flexible boot   9   Jump ring
4   CV joint

3   Driveshaft gaiter – renewal

1   If a split or hole in a driveshaft gaiter has been overlooked for any length of time, then the joint must be dismantled and thoroughly cleaned before renewing the gaiter.
2   Remove the driveshaft, as described in the preceding Section.

Outboard gaiter
3   Cut through the circular band which retains the constant velocity joint gaiter in position. Peel back the gaiter and expose the joint.
4   Hold the driveshaft upright and using a soft-faced mallet strike the edge of the constant velocity joint until it releases from the driveshaft.
5   Extract the boot from the driveshaft.

Inboard gaiter
6   Cut through the circular bands which secure the gaiter to the driveshaft and the sliding joint.
7   Extract the gaiter from the driveshaft.

Either joint
8   Wipe away as much grease as possible from the joints and from the outside of the driveshaft.

Reassembly
9   Locate the new gaiters on the driveshaft and then fit the driveshaft to the inner member of the constant velocity joint. Use a soft-faced mallet and compress the jump ring on the driveshaft to facilitate entry of the shaft into the inner member.

10 Now pack each joint with the correct quantity of lubricant which is:

Sliding (inner joint) – 150 cc of Duckhams Adgear 00 grease
CV (outer joint) – 52 cc of Duckhams Bentone Q5795

11 Secure the gaiter using new bands. Make sure that the folds in the clips are located so that they are towards the forward rotation of the driveshafts.
12 Pull the clip tight, close the front tabs and then fold the end of the clip over and close the rear tabs.
13 Refit the driveshaft as described in the preceding Section.

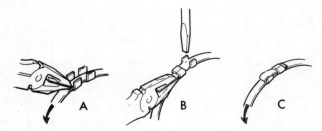

Fig. 8.5 Fitting stages of a driveshaft gaiter clip (Sec 3)

Arrow indicates forward direction of rotation of shaft

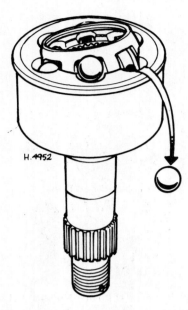

Fig. 8.6 Extracting balls from CV joint (Sec 4)

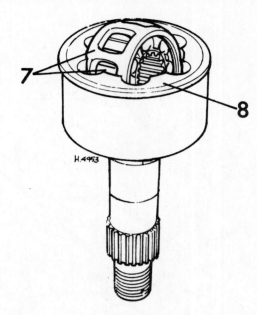

Fig. 8.7 Removing inner member from CV joint (Sec 4)

    7   *Inner member and ball cage*   8   *Outer member*

9   Pack the joint with lubricant, as described in Section 3 and then fit the driveshaft to the inner member using a soft-faced mallet.
10 Fit the gaiter clips as described in Section 3.

5   Driveshaft inboard (sliding) joint – overhaul

1   Remove the driveshaft as described in Section 2.
2   Remove the sliding joint flexible gaiter as described in Section 3.
3   Remove and discard the retaining ring from the lip of the sliding joint and withdraw the driveshaft from the sliding joint outer member. This ring is only used to prevent the joint being pulled apart during the building of the car and it can be discarded. Replacement of the ring is not necessary.
4   Remove the jump ring from the driveshaft and withdraw the sliding joint inner member and ball cage assembly.
5   Prise the balls from the ball cage.
6   Rotate the joint inner member inside the ball cage until the lands on the inner member coincide with the grooves inside the ball cage and then withdraw the inner member.
7   Clean and examine all components and renew any that are worn.
8   Reassembly is a reversal of dismantling but note that the inner member and ball cage must be fitted to the driveshaft so that the narrower outer diameter of the ball cage is towards the CV joint end of the driveshaft.
9   Lubricate the joint and fit the gaiter clips as described in Section 3.

Fig. 8.8 Removing inner member from chamfered side of CV joint ball cage (Sec 4)

   9   *Land aligned with large*   10 *Inner member*
           window

4   Driveshaft outboard (CV) joint – overhaul

1   Remove the driveshaft (Section 2).
2   Remove the CV joint flexible gaiter as described in Section 3.
3   Tilt and swivel the inner member of the joint and the ball cage in the outer member until the balls can be prised from the cage.
4   Swivel the inner member and the ball cage into alignment with the joint axis and then rotate the cage until its two large windows coincide with two of the lands in the joint outer member.
5   Withdraw the inner member and cage from the outer member.
6   Swivel the inner member into alignment with the axis of the ball cage so that two lands of the inner member coincide with the two large windows in the ball cage and then withdraw the inner member from the chamfered bore side of the ball cage.
7   Clean and examine all components and renew any that are worn.
8   Reassembly is a reversal of dismantling but note that the inner member and cage must be fitted into the outer member with the chamfered bore side of the cage at the blind end of the outer member and the lugs on the inner member at the open end of the outer member.

6   Front hub bearings – removal and refitting

1   On some models with a detachable hub cap it may be possible to release the driveshaft nut before raising the front of the car. Where this is not possible, raise the car, remove the roadwheel and release the driveshaft nut while an assistant fully applies the footbrake to prevent the hub from rotating.
2   Unbolt the caliper and tie it up out of the way to prevent strain on the flexible hoses (Fig. 8.11).
3   From the end of the driveshaft, remove the hub nut and split collar.
4   Tap the end of the driveshaft from the drive flange assembly using a soft-faced mallet or alternatively use a two or three-legged puller applying pressure from the centre screw to the end of the driveshaft.
5   Disconnect the tie-rod end balljoint from the steering arm.

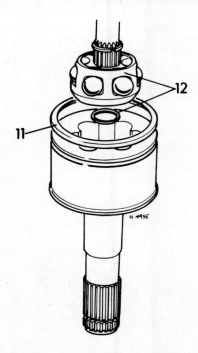

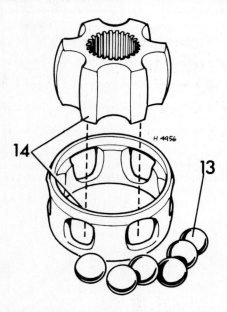

Fig. 8.10 Driveshaft sliding joint inner member and ball cage (Sec 5)

13 Balls
14 Lands and groove in alignment
to permit separation of components

Fig. 8.9 Driveshaft sliding joint (Sec 5)

11 Retaining ring (disposable)
12 Jump ring and inner member

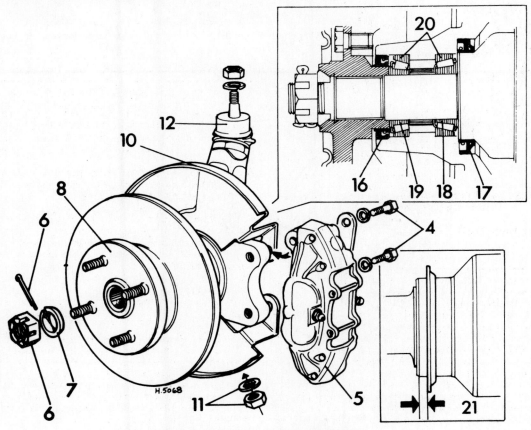

Fig. 8.11 Front hub components (Sec 6)

4 Caliper bolts	8 Hub flange	16 Inner oil seal	19 Inner bearing
5 Caliper	10 Disc shield	17 Outer oil seal	20 Bearing outer tracks
6 Hub nut	11 Lower swivel nut	18 Outer bearing	21 Location of water shield
7 Split collar	12 Upper swivel balljoint		

6   Remove the disc dust shield.

7   Disconnect the suspension lower arm from the ball pin of the swivel joint.

8   Disconnect the suspension upper arm from the ball pin of the swivel joint.

9   Withdraw the hub assembly and pull off the water shield from the driveshaft.

10   Extract the inner and outer oil seals, the bearing races and the bearing spacer.

11   Using a soft metal drift, drive out the inner and outer bearing races from the hub.

12   Refitting is a reversal of removal but if both front wheel bearings are being renewed, do not mix the components of the bearing sets as they are matched in production. Always use new oil seals and pack the bearings with fresh grease.

13   Locate a new water shield at a position $\frac{1}{4}$ in (6.0 mm) onto the hub.

14   Tighten all bolts and nuts to the specified torque.

7   Rear hub bearings – removal and refitting

1   Chock the front roadwheels, raise the rear of the car and remove the roadwheel.

2   Tap off the grease cap (photo).

3   Extract the split pin and unscrew the hub nut. Remember that the left-hand hub nut has a left-hand thread and the right-hand one a right-hand thread (photos).

4   Remove the thrust washer and then withdraw the brake drum/hub. If the drum is locked onto the brake shoes, refer to Chapter 9 for the method of releasing the automatic adjuster (photo).

5   If the race of the inboard hub bearing is found to be stuck on the hub shaft after withdrawal of the drum/hub, pull it off using a suitable extractor.

6   Extract the oil seal, drive out the inner race of the inboard bearing and remove the tubular spacer.

7   Drive out the inner race of the outboard bearing.

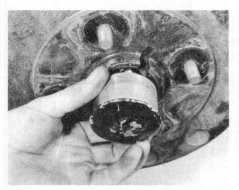

7.2 Removing rear hub grease cap

7.3A Removing rear hub split pin

7.3B Removing rear hub nut

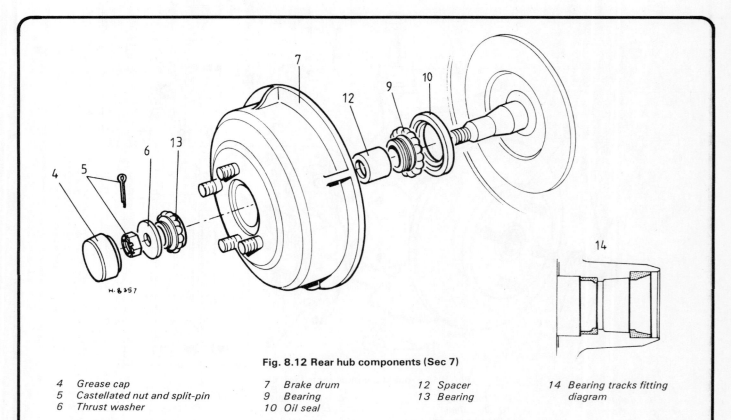

H.8257

Fig. 8.12 Rear hub components (Sec 7)

4 Grease cap	7 Brake drum	12 Spacer	14 Bearing tracks fitting
5 Castellated nut and split-pin	9 Bearing	13 Bearing	diagram
6 Thrust washer	10 Oil seal		

7.4 Rear hub thrust washer

7.9 Tightening rear hub nut

8   Drive the bearing outer tracks from the hub. If the outer track for the inboard bearing is of narrow type with a plastic spacer between the track and the oil seal, do not refit this spacer if a new bearing of this type is being installed.

9   Reassembly and refitting are reversals of the removal and dismantling operations, but observe the following:

(a) *Pack the bearings with specified grease*
(b) *The bearing tubular spacer must be fitted so that its narrower internal diameter is adjacent to the outer bearing*
(c) *The oil seal must be fitted flush with the housing*
(d) *Do not fill the hub cap with grease*
(e) *Tighten the hub nut to the specified torque. If necessary, tighten it a little more to align the split pin hole, never unscrew it to achieve alignment (photo).*

8   Roadwheels and tyres

1   Whenever the roadwheels are removed it is a good idea to clean the insides of the wheels to remove accumulations of mud and in the case of the front ones, disc pad dust.

2   Check the condition of the wheel for rust and repaint if necessary.

3   Examine the wheel stud holes. If these are tending to become elongated or the dished recesses in which the nuts seat have worn or become overcompressed, then the wheel will have to be renewed.

4   With a roadwheel removed, pick out any embedded flints from the tread and check for splits in the sidewalls or damage to the tyre carcass generally.

5   Where the depth of tread pattern is 1 mm or less, the tyre must be renewed.

6   Rotation of the roadwheels to even out wear is a worthwhile idea if the wheels have been balanced off the car. Include the spare wheel in the rotational pattern. With radial tyres it is recommended that the wheels are moved between front and rear on the same side of the car only.

7   If the wheels have been balanced on the car then they cannot be moved round the car as the balance of wheel, tyre and hub will be upset. In fact their exact stud fitting positions must be marked before removing a roadwheel so that it can be returned to its original 'in balance' state.

8   It is recommended that wheels are re-balanced halfway through

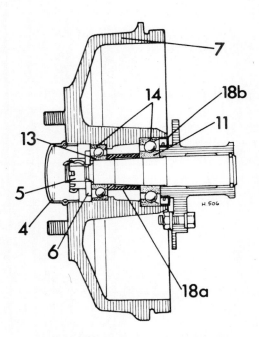

Fig. 8.13 Sectional view of rear hub (Sec 7)

4	Dust cap	13	Outer bearing
5	Hub nut	14	Bearing outer tracks
6	Special washer	18a	Bearing spacer
7	Brake drum	18b	Oil seal
11	Inner bearing		

the life of the tyres to compensate for the loss of tread rubber due to wear.

9   Finally, always keep the tyres (including the spare) inflated to the recommended pressures and always replace the dust caps on the tyre valves. Tyre pressures are best checked first thing in the morning when the tyres are cold.

9   Fault diagnosis – driveshafts

Symptom	Reason(s)
Vibration	Driveshafts bent Driveshafts out of balance Worn or 'dry' joints Roadwheels/tyres need balancing
'Clunk' on taking up drive or on deceleration	Worn driveshaft splines Loose hub nut Loose roadwheel nuts
Noise or roar especially when cornering	Worn hub bearings Incorrectly adjusted hub bearings

Chapter 9 Braking system

Contents

Specifications

System type	Four wheel, dual-circuit, hydraulic with servo assistance. Discs front, drums rear. Handbrake mechanical to rear wheels

Hydraulic fluid
Type/specification .... Hydraulic fluid to FMVSS DOT 3 (Duckhams Universal Brake and Clutch Fluid)

Discs
Diameter .... 10.64 in (270.3 mm)
Wear limit of pad friction material ..... 0.12 in (3.0 mm)

Drums
Diameter .... 9.0 in (228.6 mm)
Wear limit of shoe friction material ..... 0.06 in (1.5 mm)

Master cylinder bore diameter .... 0.813 in (20.65 mm)

Rear wheel cylinder bore diameter ..... 0.563 in (14.3 mm)

Vacuum servo
Type .... Lockheed 65LR direct acting

Torque wrench settings

	lbf ft	Nm
Caliper mounting bolts	55	75
Disc dust shield bolts	15	20
Disc to drive flange bolts	45	61
Master cylinder mounting nuts	17	23
Rear hub nut	60	82
Rear backplate nuts	20	27
Bleed screws	7	10
Brake pedal bracket to servo	15	20
Steering column to brackets	15	20
Steering intermediate shaft to column	15	20
Flexible hose to caliper	9	12
Handbrake screws to floor	15	20
Pressure regulating valve end plug	40	55
Pressure regulating valve fixing screw	9	12

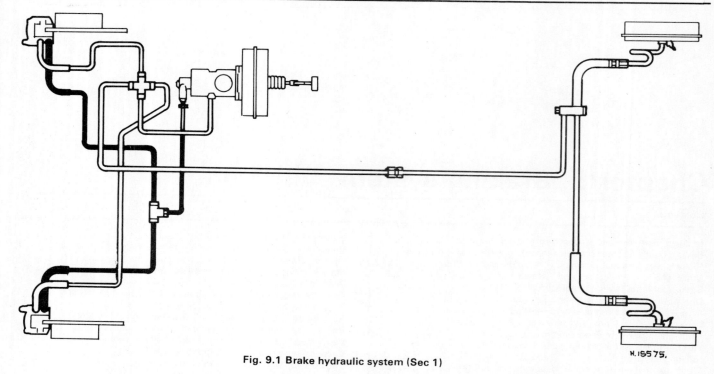

Fig. 9.1 Brake hydraulic system (Sec 1)

Primary circuit (parallel lines) includes rear wheel cylinders and front caliper lower cylinders
Secondary circuit (solid line) includes front caliper upper cylinders

1 Description and maintenance

1   The braking system is of dual circuit hydraulic type with discs at the front and drums at the rear.
2   The two hydraulic circuits are arranged in the following way. The primary circuit is the two lower cylinders of both front calipers plus the two rear wheel cylinders. The secondary circuit is the two upper cylinders of both front calipers.
3   A pressure regulating valve is incorporated in the hydraulic circuit to limit pressure to the rear wheel cylinders in order to prevent the rear wheels locking during heavy brake applications.
4   A vacuum servo unit is fitted to all models.
5   Warning lamps are provided for brake pad wear and low hydraulic fluid level.
6   The handbrake operates through mechanical linkage to the rear brake shoes which are of self-adjusting type.
7   Check the fluid level, pad and lining wear at the intervals specified in Routine Maintenance (photo).

2 Disc pads – inspection and renewal

1   Although the disc pads incorporate wear sensors, it is recommended that the pads are inspected at the intervals specified in Routine Maintenance in order to ensure maximum braking efficiency.
2   Raise the front of the car and removed the roadwheels. Inspect the thickness of the friction material. If it is less than $\frac{1}{8}$ in (3.0 mm) then the pads must be renewed as an axle set.
3   Disconnect the pad wear sensor leads and withdraw the pad retaining pins (photos).
4   Remove the anti-rattle springs and then grip each of the pads in turn with a pair of pliers and extract them from the caliper (photos).
5   Brush out any accumulated dust and dirt.
6   In order to accommodate the increased thickness of the new pads, the caliper pistons should be depressed evenly and equally into their cylinders using a short piece of wood or flat metal bar. As the pistons are pressed in, the fluid level in the master cylinder will rise and fluid will be displaced so either draw off some fluid from the reservoir using an old clean hydrometer or alternatively fit a bleed tube to the caliper bleed nipple and submerge its open end in a jar containing a little

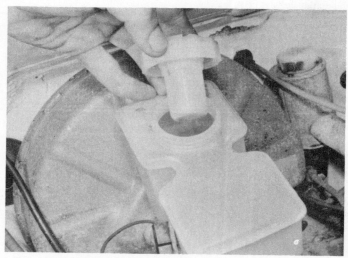

1.7 Checking fluid level

hydraulic fluid. Release the bleed nipple and as the pistons are depressed, fluid will be expelled into the jar. Retighten the bleed nipple.
7   Fit the new disc pads making sure that the friction surface is against the disc. No anti-squeal shims are fitted.
8   Refit the anti-rattle springs (longer legs to centre) and push in the new retaining pins supplied with the new pads. Bend up the ends of the pins.
9   Apply the footbrake hard several times to bring the pads into contact with the discs, refit the roadwheels and lower the car.
10 Check and top-up the master cylinder reservoir to the correct level using clean fluid. Discard any fluid which was bled from the system earlier.
11 Always fit new disc pads in complete axle sets – never renew one side only.

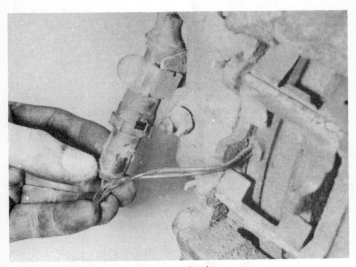

2.3A Disconnecting pad wear sensor leads

2.3B Removing pad retaining pin

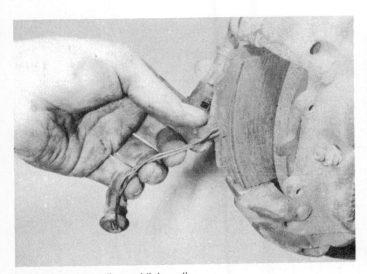

2.4A Removing a disc pad (inboard)

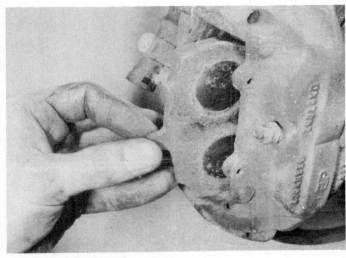

2.4B Removing a disc pad (outboard)

3   Brake shoes – inspection and renewal

1   At the intervals specified in Routine Maintenance, slacken the rear roadwheel nuts, release the handbrake and jack-up the rear of the car. Support the car under the rear suspension body bracket.

2   Remove the roadwheel and withdraw the grease retaining cap.

3   Extract the split pin and retainer from the hub nut. Unscrew and remove the nut and special washer. On the left-hand side of the car, the nut has a left-hand thread and on the right-hand side of the car it has a right-hand thread.

4   Pull the hub/drum assembly from the stub axle. If it is tight, tap it off gently using a soft-faced mallet or employ a suitable extractor.

5   After very high mileages or if the interior of the drum has become grooved (due to failure to renew linings before the rivets contact the friction surface), then it is possible that the action of the automatic adjuster may cause the shoes to lock the drum onto the axle and prevent its removal. In these circumstances, access to the automatic adjuster will have to be obtained by drilling a hole in the brake drums. The hole should be drilled in accordance with the diagram (Fig. 9.2). Now turn the drum until the hole is level with the centre of the hub and nearer the rear of the car, then insert a thin screwdriver and lift the release lever on the automatic adjuster mechanism. When reassembling, seal the hole with a rubber plug.

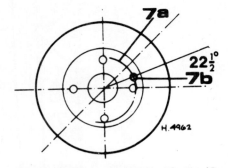

Fig. 9.2 Brake drum drilling diagram (Sec 3)

7a Drilling arc   *7b 8.0 mm hole*

6   With the brake drum removed, brush away all dust from the shoes and drum insulator, taking care not to inhale the dust. If the linings are in good condition, the drum can be refitted but renew the shoes if they are worn down or nearly down to the rivets. Bonded lining minimum material thickness is $\frac{1}{16}$ in (1.5 mm). Renew the shoes on an exchange basis, it is not worth attempting to reline shoes yourself. If there is any evidence of oil leakage, this must be rectified before fitting the new

3.6 Left-hand drum brake

1 Leading shoe
2 Trailing shoe

3.7 Releasing washer from shoe steady pin

shoes and it will probably be due to a faulty wheel cylinder (see Section 6) or hub oil seal (see Chapter 8) (photo).
7   To renew the shoes, release the shoe steady spring and dished washer from the trailing shoe. To do this, grip the slotted dished washer with a pair of pliers, depress it and turn it through 90°. It can then be withdrawn from the T-shaped head of the steady post and the spring removed (photo).
8   Ease the trailing shoe from its anchorage and extract the shoe upper return spring.
9   Disengage and withdraw the cross lever and its spring.

10 Use a pair of pliers to release the lower return spring and remove the trailing shoe.
11 Remove the steady spring and washer from the leading shoe, turn the shoe so that the handbrake can be released from its lever and then withdraw the leading shoe.
12 Place an elastic band or a piece of wire round the wheel cylinder pistons to prevent them falling out and do not touch the footbrake pedal while the shoes and drum are withdrawn.
13 Lay the new shoes on the bench in their correct fitted attitude noting carefully the relative positions of the leading and trailing ends

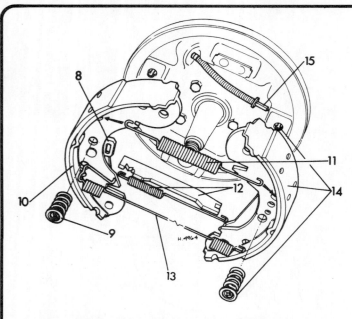

Fig. 9.3 Rear brake assembly (LH side) (Sec 3)

8   Adjuster lever
9   Shoe steady dished washer
10 Leading shoe
11 Upper shoe return spring
12 Cross lever and spring
13 Lower return spring
14 Trailing shoe and steady post and spring
15 Handbrake cable

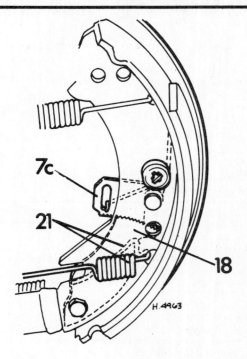

Fig. 9.4 Rear brake shoe automatic adjuster lever (7c) adjuster mechanism (18) and lower return spring (21) (Sec 3)

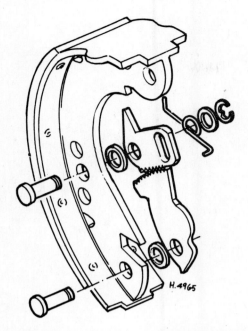

Fig. 9.5 Leading shoe and components of the automatic adjuster (Sec 3)

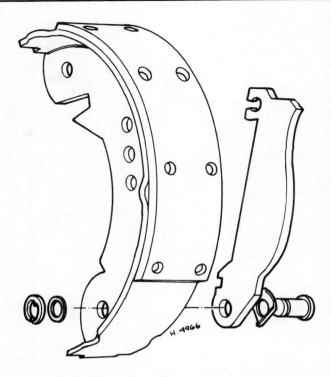

Fig. 9.6 Trailing shoe and handbrake lever (Sec 3)

of the shoes (the greater or smaller areas of exposed web not covered by the friction lining).

14 Transfer the automatic adjuster mechanism to the inside face of the new trailing shoe. Fit the pivot pins after applying a smear of high melting point grease to them.

15 Set the mechanism to the minimum adjustment position by lifting the release lever and moving the adjuster lever towards the friction lining.

16 Transfer the handbrake lever to the inside face of the leading shoe. Smear the pivot pin with high melting point grease and make sure that the spring washer has its concave side to the lever.

17 Apply a smear of high melting point grease to the shoe contact points of the brake backplate, the wheel cylinder and lower shoe anchorage and the cross lever.

18 Fit the lower return spring to the trailing shoe making sure that the inside end is over the automatic adjuster lever.

19 Fit the shoes, the cross lever (making sure that its return spring is below the lower shoe return spring), the upper return spring and the steady springs and dished washers.

20 Refit the brake drum, tightening the hub nut to the specified torque. If necessary, in order to insert the split pin, overtighten the nut slightly, never back it off.

21 Fit the roadwheel and then repeat the operations on the opposite wheel. Never renew one set of brake shoes only, always renew them as complete axle sets.

22 When the car has been lowered to the ground, apply the footbrake several times to operate the automatic adjuster mechanism and then check and top-up, if necessary, the master cylinder reservoir.

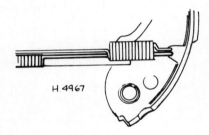

Fig. 9.7 Correct location of cross lever spring underneath shoe lower return spring (Sec 3)

4.4 Withdrawing caliper unit

---

4 Front disc caliper – removal, overhaul and refitting

1 Jack-up the front of the car, support it securely and remove the roadwheel.

2 Extract the disc pads (Section 2).

3 Disconnect the hydraulic hoses making sure that they are identified for correct reconnection. Unscrew the union on the rigid pipelines and plug or cap their ends to prevent loss of fluid. Then unscrew and remove the nuts and lockwashers from the ends of the flexible hoses to release them from their body support brackets.

4 Unscrew and remove the two bolts which secure the caliper to the stub axle carrier. Remove the caliper and unscrew the flexible hoses from it (photo).

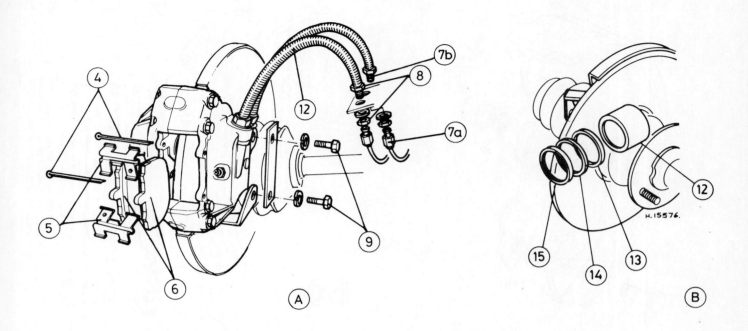

Fig. 9.8 Disc brake components (Sec 4)

A		B	
4 Pad pins	7b Flexible hose nut and lockwasher	12 Piston	14 Boot retainer
5 Anti-rattle clips	8 Secondary hose and union	13 Piston seal	15 Dust excluding boot
6 Pads	9 Caliper securing bolts		
7a Fluid line union	12 Flexible brake hose		

5   Clean all dirt from the external surfaces of the unit and drain any brake fluid from it.

6   Place a thin piece of wood between the pistons and depress two pistons on the same side with the fingers. Now apply air pressure from a foot pump at one of the fluid entry holes while placing a finger over the other hole. One of the pistons will be ejected and can be removed.

7   Apply air pressure at the other fluid entry hole, placing the finger over the hole previously used, and eject the second piston. Identify both pistons in respect of their cylinders by using a piece of masking tape. Clean components only with methylated spirit or hydraulic fluid – nothing else.

8   From the two empty cylinders, prise the wiper seal and retainer and the piston seal, use a blunt probe to prevent scratching the cylinder bores. Should either piston or cylinder surface show signs of scoring or 'bright' wear areas, then the caliper must be renewed complete.

9   Obtain the appropriate repair kit and fit the new seals and retainer using the fingers only to manipulate them into position.

10   Dip each piston in brake fluid and insert it squarely into its original cylinder bore.

11   Holding the two pistons, just serviced, in the depressed position, eject the other two pistons in a similar manner to that just described. Fit new seals and inspect the piston and cylinder surfaces for scoring or 'bright' wear areas, renewing the complete unit if evident.

12   Refitting is a reversal of removal but on completion, bleed the hydraulic circuit, as described in Section 12.

---

5   Disc – inspection, removal and refitting

1   When a considerable mileage has been covered, or if the disc pads have not been renewed and have worn below the specified minimum thickness, check the disc for distortion or run-out and grooving.

2   Light scoring or grooving is normal and should be ignored. Heavy scoring or deep grooving will necessitate renewal of the disc.

3   Jack up the front of the car and remove the roadwheel and disc pads.

4   Using either a dial gauge or feeler blades against a fixed point, turn the hub/disc assembly and check that any run-out does not exceed 0.008 in (0.2 mm) otherwise the disc will have to be renewed.

5   To remove the disc, refit the pads and have an assistant apply the brakes hard while the split pin is extracted and the driveshaft nut unscrewed.

6   Remove the caliper, as described in Section 4, and tie it up out of the way to avoid straining the hoses.

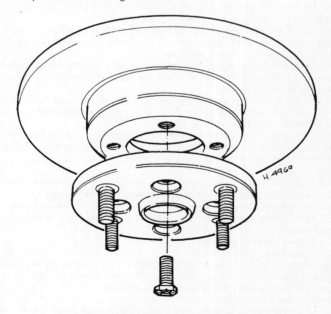

Fig. 9.9 Brake disc and drive flange (Sec 5)

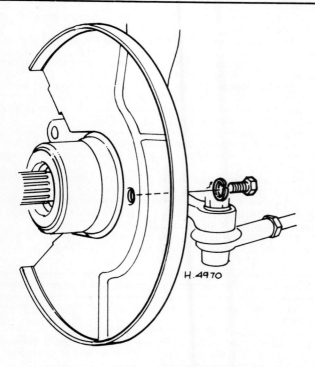

Fig. 9.10 Disc dust shield and securing bolt (Sec 5)

7   Remove the driveshaft nut and split collar.
8   Pull the driving flange/disc assembly from the driveshaft. If it is tight, use a suitable puller.
9   Grip the disc in a vice fitted with jaw protectors and remove the bolts which secure the disc to the driving flange.
10   If required, the disc brake shield can be removed after extracting the two securing bolts.
11   Refitting is a reversal of removal, but tighten all bolts and nuts to the specified torque.

6   Rear wheel cylinder – removal, overhaul and refitting

1   Raise the rear of the car and remove the roadwheel. Release the handbrake.
2   Tap off the grease cap from the hub.
3   Extract the split pin, unscrew the hub nut and remove the special washer. Remember that the left-hand nut has a left-hand thread while the right-hand one has a right-hand thread.
4   Pull the hub/drum from the stub axle. If it is tight, use a copper or plastic-faced hammer or a three-legged puller. If the drum has become grooved by the friction of the brake shoes, release the automatic adjuster by referring to Section 3 of this Chapter.
5   Refer to Section 3 and remove the steady spring from the trailing shoe.
6   Prise the end of the trailing shoe from the slot in the wheel cylinder piston.
7   Disengage and remove the cross-lever and the spring.
8   Remove the upper shoe return spring.
9   Disconnect the lower shoe return spring from the leading shoe and then lift away the trailing shoe assembly.
10   Retain the piston in the wheel cylinder with an elastic band or a piece of wire. Do not touch the brake pedal from this point on.
11   Unscrew the union nut and disconnect the fluid pipe from the wheel cylinder. Cap the end of the pipe to prevent loss of fluid.
12   Prise out the clip which retains the cylinder to the brake backplate (photo).
13   Remove the cylinder and the joint gasket.
14   Clean the external surfaces of the cylinder and then peel off the dust cover from each end of the cylinder body.
15   Withdraw the pistons and the coil spring.
16   Discard the rubber seals and dust covers, wash the components in methylated spirit or clean hydraulic fluid and examine the surfaces of the pistons and cylinder bores for scoring or 'bright' wear areas. If these are evident, renew the wheel cylinder complete.
17   If the components are in good condition, obtain a repair kit and fit new seals, using the fingers only to manipulate them into position. Dip each piston/seal assembly in clean hydraulic fluid and enter it into the cylinder body making sure that the coil spring is located between the pistons.

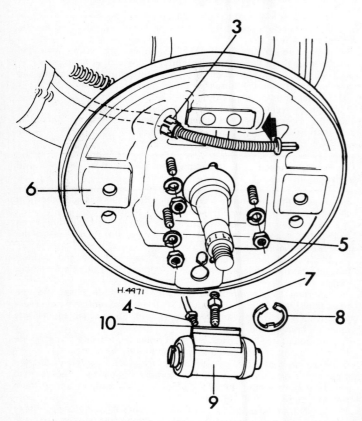

Fig. 9.11 Rear brake backplate attachments (Sec 6)

3	Handbrake cable clip	8	Wheel cylinder retaining
4	Fluid pipe union		circlip
5	Nut	9	Wheel cylinder
6	Backplate	10	Gasket
7	Bleed screw		

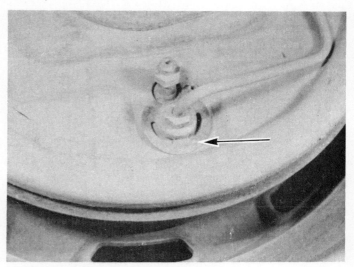

6.12 Rear wheel cylinder fixing clip (arrowed)

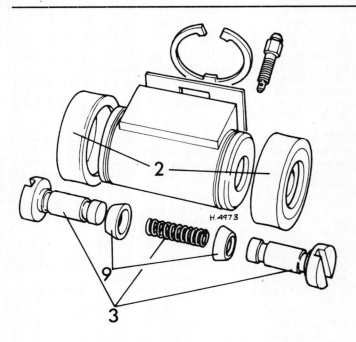

Fig. 9.12 Exploded view of a rear wheel cylinder (Sec 6)

2   Dust excluding boots   9   Piston seals
3   Piston and spring

18 Refit the wheel cylinder and shoe components by reversing the removal operations and reference to Section 3.
19 Refit the drum and tighten the nut to the specified torque.
20 Bleed the hydraulic circuit as described in Section 12.

7   Brake drum – removal, inspection and refitting

1   Removal of the rear brake drum is described in Section 3.
2   With the drum removed, inspect the interior friction surface for grooves. If these are evident, renew the drum. Internal calipers will be required to check for ovality, a condition which can occur after a high mileage and must be rectified by renewal of the drum.
3   Refitting is as described in Section 3.

8   Rear brake backplate – removal and refitting

1   To remove the rear brake backplate, first withdraw the shoe assembly (see Section 3).
2   Compress the clip which retains the handbrake cable to the backplate and then draw the cable through the backplate.
3   Disconnect the hydraulic brake line from the wheel cylinder and cap the line to prevent loss of fluid.
4   Remove the three bolts which retain the backplate to the radius arm and withdraw the backplate.
5   The wheel cylinder can be removed after extracting the securing circlip.
6   Refitting is a reversal of removal but remember to bleed the hydraulic circuit, as described in Section 12.

9   Master cylinder – removal, overhaul and refitting

1   Syphon as much fluid as possible from the master cylinder reservoir.
2   Disconnect the electrical leads from the reservoir cap.
3   Unscrew the union nuts and disconnect the pipelines from the master cylinder. Be prepared to catch leaking fluid with pads of rag. Cap the ends of the pipes.

4   Unscrew the nuts and withdraw the master cylinder from the vacuum servo unit.
5   Clean away external dirt and then 'rock' the reservoir out of its sealing rings.
6   Depress the primary piston with a rod and extract the circlip. Extract the primary piston and spring.
7   Depress the secondary piston and extract the stop pin which is accessible from within the reservoir forward seal recess.
8   Withdraw the secondary piston and return spring. Use air from a tyre pump if necessary to eject them.
9   Wash all components in methylated spirit or clean hydraulic fluid and examine the surfaces of the pistons and cylinder bore for scoring or 'bright' wear areas. Where these are evident renew the complete master cylinder.
10 If the components are in good condition, discard all seals and obtain the appropriate repair kit.
11 Manipulate the new seals into position using the fingers only.
12 Dip the assembled pistons in clean hydraulic fluid before re-assembling, which is the reverse of dismantling.
13 Bolt the master cylinder into position, reconnect the fluid pipes and then bleed the entire hydraulic system as described in Section 12.

10 Pressure regulating valve – removal, overhaul and refitting

1   Raise the rear of the car and remove the right-hand rear roadwheel.
2   Disconnect and cap the fluid pipes from the valve.
3   Disconnect the flexible hose from the bracket on the radius arm.
4   Unscrew the valve fixing bolt and withdraw the valve until the flexible hose can be uncoupled.
5   Clean away external dirt and then grip the valve in the jaws of a vice.
6   Unscrew the endplug and extract the internal components.
7   If the cylinder bore of the valve is scored or corroded, renew the valve complete.
8   If the components are in good condition, discard the seals and obtain a repair kit which will contain all the essential new items.
9   Manipulate the new seals into position using the fingers only, dip the components in clean hydraulic fluid and reassemble.
10 Tighten the end plug to the specified torque.
11 Refit the valve, connect the fluid lines.
12 Refit the roadwheel and lower the car to the floor.
13 Bleed the complete hydraulic system.

11 Flexible hoses and rigid pipelines – inspection, removal and refitting

1   Inspect the condition of the flexible hydraulic hoses. If they are swollen, perished or chafed, they must be renewed.
2   To remove a flexible hose, hold the flats on its end-fitting in an open ended spanner and unscrew the union nut which couples it to the rigid brake line (photo).
3   Disconnect the flexible hose from the rigid line and support bracket and then unscrew the hose from the caliper or wheel cylinder circuit as the case may be.
4   Refitting is a reversal of removal. The flexible hoses may be twisted not more than one quarter turn in either direction if necessary to provide a 'set' to ensure that they do not rub or chafe against any adjacent component.
5   At regular intervals wipe the steel brake pipes clean and examine them for signs of rust or denting caused by flying stones.
6   Examine the securing clips which are plastic coated to prevent wear to the pipe surface. Bend the tongues of the clips if necessary to ensure that they hold the brake pipes securely without letting them rattle or vibrate.
7   Check that the pipes are not touching any adjacent components or rubbing against any part of the vehicle. Where this is observed bend the pipe gently away to clear.
8   Although the pipes are plated any section of pipe may become rusty through chafing and should be renewed. Brake pipes are available to the correct length and fitted with end unions from most Austin-Rover dealers and can be made to pattern by many accessory suppliers. When fitting the new pipes use the old pipes as a guide to bending and do not make any bends sharper than is necessary.

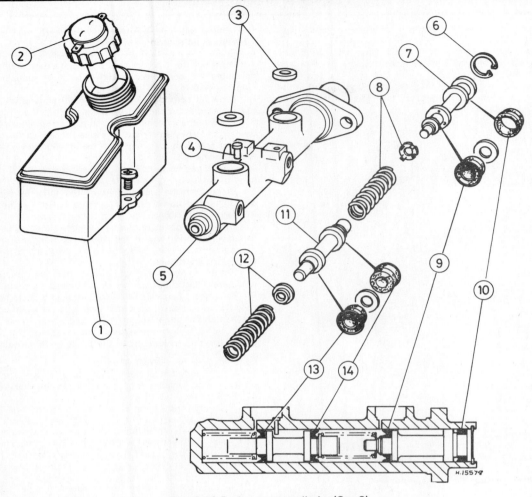

Fig. 9.13 Brake master cylinder (Sec 9)

1 Fluid reservoir
2 Filler cap and level sensor
3 Sealing rings
4 Secondary piston stop pin
5 Master cylinder body
6 Circlip
7 Primary piston
8 Return spring and cup
9 Seal and washer
10 Seal
11 Secondary piston
12 Return spring and cup
13 Seal and washer
14 Seal

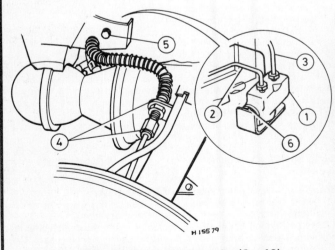

Fig. 9.14 Pressure regulating valve (Sec 10)

1 Pressure valve
2 Inlet pipe
3 Outlet pipe
4 Outlet hose
5 Valve mounting screw
6 Valve end plug

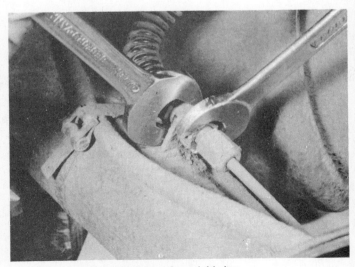

11.2 Disconnecting flexible hose from rigid pipe

9   The system will of course have to be bled when the circuit has been reconnected.
10 Brake fittings may be encountered which have UNF or metric threads. Always test for compatibility by screwing the hose or end fitting in by hand initially.
11 It should be noted that a metric hose seals against the bottom of the port. Never attempt to eliminate the gap (arrowed) by overtightening the end fitting.

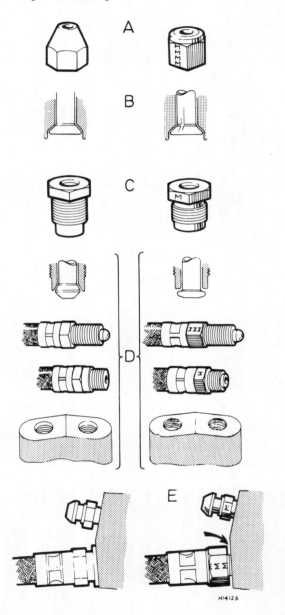

Fig. 9.15 Unified and Metric fittings used in the braking system (Sec 11)

A – *Metric pipe unions, bleed screws and hose ends are coloured black or gold. Most are also identified by the letter M.*
B – *The correct Unified or Metric pipe flares must be used.*
C – *The end of a Metric hose is coloured black or gold.*
D – *Metric fittings are not counterbored. Some Unified fittings may also not be counterbored. If the thread type is not known on a fitting, screw the item in by finger pressure only. If the fit is slack, or the item will not screw fully in, the threads may be of different types.*
E – *A Metric hose seals against the bottom of the port, with a gap between the cylinder or caliper and the hose hexagon.*

12 Hydraulic system – bleeding

1   If the master cylinder or the pressure regulating valve have been disconnected and reconnected then the complete system (both circuits) must be bled.
2   If only a component of one circuit has been disturbed then only that particular circuit need be bled.
3   If the entire system is being bled, the sequence of bleeding should be carried out by starting at the bleed screw furthest from the master cylinder and then following with the bleed screw on the opposite rear wheel cylinder.
4   Now bleed from the two lower screws on one front caliper (do this simultaneously).
5   Bleed from the upper screw on the same caliper just mentioned.
6   Bleed both lower screws, simultaneously on the opposite caliper followed by the upper screw on that caliper.
7   Unless the pressure bleeding method is being used, do not forget to keep the fluid level in the master cylinder reservoir topped up to prevent air from being drawn into the system which would make any work done worthless.
8   Before commencing operations, check that all system hoses and pipes are in good condition with all unions tight and free from leaks.
9   Take great care not to allow hydraulic fluid to come into contact with the vehicle paintwork as it is an effective paint stripper. Wash off any spilled fluid immediately with cold water.
10 If the system incorporates a vacuum servo, destroy the vacuum by giving several applications of the brake pedal in quick succession.

Bleeding – two-man method
11 Gather together a clean glass jar and two lengths of rubber or plastic tubing which will be a tight fit on the brake bleed screws.
12 Engage the help of an assistant.
13 Push one end of the bleed tube onto the first bleed screw and immerse the other end in the glass jar which should contain enough hydraulic fluid to cover the end of the tube (photo).
14 Open the bleed screw one half a turn and have your assistant depress the brake pedal fully then slowly release it. Tighten the bleed screw at the end of each pedal downstroke to obviate any chance of air or fluid being drawn back into the system.
15 Repeat this operation until clean hydraulic fluid, free from air bubbles, can be seen coming through into the jar.
16 Tighten the bleed screw at the end of a pedal downstroke and remove the bleed tube. Bleed the remaining screws in a similar way.

Bleeding – using one-way valve kit
17 There is a number of one-man, one-way, brake bleeding kits available from motor accessory shops. It is recommended that one of these kits is used wherever possible as it will greatly simplify the bleeding operation and also reduce the risk of air or fluid being drawn back into the system quite apart from being able to do the work without the help of an assistant.
18 To use the kit, connect the tube to the bleed screw and open the screw one half a turn.
19 Depress the brake pedal fully and slowly release it. The one-way valve in the kit will prevent expelled air from returning at the end of each pedal downstroke. Repeat this operation several times to be sure of ejecting all air from the system. Some kits include a translucent container which can be positioned so that the air bubbles can actually be seen being ejected from the system.
20 Tighten the bleed screw, remove the tube and repeat the operations on the remaining brakes.
21 On completion, depress the brake pedal. If it still feels spongy repeat the bleeding operations as air must still be trapped in the system.

Bleeding – using a pressure bleeding kit
22 These kits too are available from motor accessory shops and are usually operated by air pressure from the spare tyre.
23 By connecting a pressurised container to the master cylinder fluid reservoir, bleeding is then carried out by simply opening each bleed screw in turn and allowing the fluid to run out, rather like turning on a tap, until no air is visible in the expelled fluid.
24 By using this method, the large reserve of hydraulic fluid provides a safeguard against air being drawn into the master cylinder during bleeding which often occurs if the fluid level in the reservoir is not maintained.

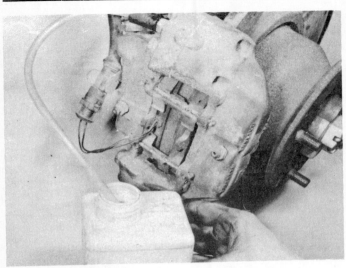

12.14 Bleeding a caliper

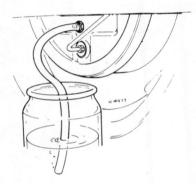

Fig. 9.16 Bleeding a rear wheel cylinder (Sec 12)

25 Pressure bleeding is particularly effective when bleeding 'difficult' systems or when bleeding the complete system at time of routine fluid renewal.

All methods
26 When bleeding is completed, check and top up the fluid level in the master cylinder reservoir.
27 Check the feel of the brake pedal. If it feels at all spongy, air must still be present in the system and further bleeding is indicated. Failure to bleed satisfactorily after a reasonable repetition of the bleeding operations may be due to worn master cylinder seals.
28 Discard brake fluid which has been expelled. It is almost certain to be contaminated with moisture, air and dirt making it unsuitable for further use. Clean fluid should always be stored in an airtight container as it absorbs moisture readily (hygroscopic) which lowers its boiling point and could affect braking performance under severe conditions.

13 Vacuum servo unit – description and testing

1 A vacuum servo unit is fitted into the brake hydraulic circuit in series with the master cylinder, to provide assistance to the driver when the brake pedal is depressed. This reduces the effort required by the driver to operate the brakes under all braking conditions (photo).
2 The unit operates by vacuum obtained from the induction manifold and consists of basically a booster diaphragm and non-return valve.

The servo unit and hydraulic master cylinder are connected together so that the servo unit piston rod acts as the master cylinder pushrod. The driver's braking effort is transmitted through another pushrod to the servo piston and its built-in control system. The servo unit piston does not fit tightly into the cylinder, but has a strong diaphragm to keep its edges in constant contact with the cylinder wall, so assuring an air tight seal between the two parts. The forward chamber is held under vacuum conditions created in the inlet manifold of the engine and, during periods when the brake pedal is not in use, the controls open a passage to the rear chamber so placing it under vacuum conditions as well. When the brake pedal is depressed, the vacuum passage to the rear chamber is cut off and the chamber opened to atmospheric pressure. The consequent rush of air pushes the servo piston forward in the vacuum chamber and operates the main pushrod to the master cylinder.
3 The controls are designed so that assistance is given under all conditions and, when the brakes are not required, vacuum in the rear chamber is established when the brake pedal is released. All air from the atmosphere entering the rear chamber is passed through a small air filter.
4 Under normal operating conditions the vacuum servo unit is very reliable and does not require overhaul except at very high mileages. In this case it is far better to obtain a service exchange unit, rather than repair the original unit.
5 It is emphasised, that the servo unit assists in reducing the braking effort required at the foot pedal and in the event of its failure, the hydraulic braking system is in no way affected except that the need for higher pedal pressures will be noticed.
6 If it is suspected that the servo unit is faulty, it can be tested in the following way. Run the engine.
7 Switch off the engine and depress the brake pedal several times to destroy the servo vacuum.
8 Now apply a firm consistent pressure to the brake pedal and start the engine. If the pedal drops slightly then the servo is operating correctly. If the pedal does not move then check the vacuum hose for a leak. The condition may also be due to a clogged air filter or faulty non-return valve (refer to the next two Sections).
9 If these items appear to be satisfactory then the servo unit itself must be faulty.

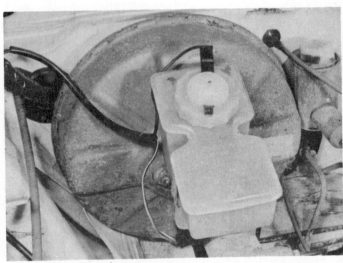

13.1 Master cylinder/servo/clutch master cylinder arrangement

14 Vacuum servo air filter – renewal

1 At the intervals specified in Routine Maintenance, pull the boot from around the brake pushrod.
2 Withdraw the retainer and pull out the filter. If this is in one piece, it can be cut from the pushrod.

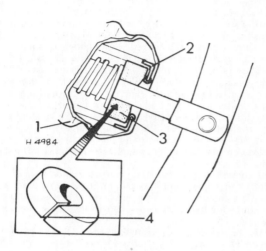

Fig. 9.17 Vacuum servo air filter (Sec 14)

1   Dust excluding boot   *3   Filter*
2   Retainer   *4   Cut*

3   To save disconnecting the pushrod, cut the new filter as shown in Fig. 9.17 and fit it. Refit the retainer and the boot.

15 Vacuum servo hose and non-return valve – renewal

1   Disconnect the vacuum hose from both the servo unit and the inlet manifold and remove it.
2   Observe the angle of the nozzle of the non-return valve and then pull the valve from its sealing grommet. A flat blade inserted between the valve and the grommet wil help in levering the valve out.
3   Pull the grommet from the servo housing gripping it securely with a pair of pliers so that it does not fall inside the unit.
4   Refitting is a reversal of removal but apply a little rubber grease or brake fluid to facilitate entry of the valve into the grommet.

16 Vacuum servo unit – removal and refitting

1   Disconnect the leads from the low fluid level warning lamp switch on the reservoir cap.
2   Disconnect the vacuum hose from the servo unit.
3   Unscrew the master cylinder mounting nuts.
4   Slacken the primary fluid feed pipe union at the four-way connector so that the master cylinder can be pulled carefully off the mounting studs. Support the cylinder to prevent bending the pipelines.
5   Disconnect the push-rod from the brake pedal.
6   Unscrew and remove the four nuts which hold the servo unit to the brake pedal bracket.
7   Working within the engine compartment, remove the vacuum servo unit from the bulkhead.
8   Refitting is a reversal of removal, but the complete hydraulic system may need bleeding if air has entered at the released union.

17 Handbrake – adjustment

1   This is not a routine operation as the rear brakes, having automatic adjusters, will normally keep the handbrake fully adjusted automatically.
2   Where there is excessive movement of the handbrake lever as well as excessive free-movement of the footbrake pedal, then faulty operation of the self-adjusting mechanism must be suspected. Where the mechanism is in good order, then the handbrake cable has probably stretched and the following adjustment should be carried out.

Excessive travel of the handbrake lever can damage the handbrake warning light switch.
3   Remove the rear seat (Chapter 12).
4   Set the handbrake lever three notches ('clicks') from the fully off position.
5   Raise the rear of the car and support it under the rear side jacking points.
6   Turn each rear roadwheel. It should be just possible to turn them against the binding action of the brakes.
7   If they turn freely without any indication of binding then the cable is slack and must be adjusted at the equaliser by turning the adjuster nut after releasing the locknut (Fig. 9.18).
8   When the binding condition has been obtained, fully release the handbrake lever and check that the rear roadwheels rotate freely without any tendency to bind.
9   Tighten the locknut to the equaliser and refit the rear seat.

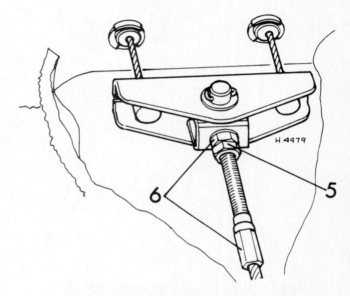

Fig. 9.18 Handbrake cable equaliser (Sec 17)

5   Locknut   *6   Adjuster nut*

18 Handbrake cables – renewal

1   Remove the front passenger seat and the rear seat (Chapter 12).
2   If the centre console is fitted, remove it (Chapter 12).
3   Remove the front seat belt anchorages from between the front seats.
4   Remove the rear door seal and carpet retainer.
5   Remove the two rear screws which secure the front door seal and carpet retainer. Pull the carpet from under the front retainer and the seat belt reel. Lift the carpet and prop it against the driver's seat.
6   Drill out the pop rivets from the guide plate and remove the plate.
7   Release the locknut which secures the front cable and disconnect the trunnion of the equaliser from the cable threaded portion.
8   Remove the bolts which secure the handbrake lever to the floor.
9   Remove the clevis pin which secures the clevis fork to the handbrake lever.
10   Extract the front handbrake cable from between the guide plates and the floor.
11   To remove a rear handbrake cable, first remove the rear brake drum (Section 3) and shoe assembly.
12   Remove the clip which secures the cable to the radius arm.
13   Pull the cable out of the retaining clip on the suspension cross tube.

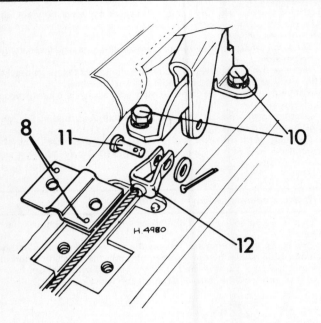

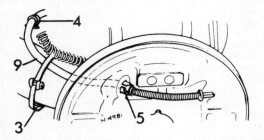

Fig. 9.20 Handbrake rear cable at backplate (Sec 18)

3   Radius arm clip   5   Backplate clip
4   Cross tube clip   9   Handbrake rear cable

Fig. 9.19 Handbrake front cable attachment to lever (Sec 18)

8   Guide plate rivets   11   Clevis pin
10   Handbrake lever bolts   12   Clevis fork

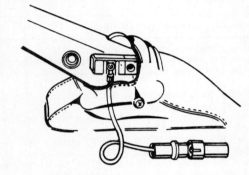

Fig. 9.21 Handbrake warning light switch (Sec 19)

14 Compress the retaining clip and pull the handbrake cable through the bracket backplate.

15 Remove the rear seat and then holding the balance lever, pull the inner cable from the slot in the equaliser.

16 Extract the clip from the outer cable and then withdraw the complete cable assembly from the car.

17 Refitting of the cable is a reversal of removal but apply grease to the clevis pins and adjust on completion, as described in the preceding Section.

18 The guide plate should be refitted using new pop rivets or small self-tapping screws.

19 Handbrake warning light switch – removal and refitting

1   If one is fitted, remove the rear centre console. To do this, prise the cover plate from its recess in the top of the console, extract the four console securing screws and then raise the rear of the console and disconnect the cigar lighter leads. Slide the console up the handbrake control lever and seat belt anchorage stalks.

2   Roll the flexible gaiter which surrounds the handbrake lever inwards and disconnect the switch wiring plug.

3   Unscrew the single retaining screw and remove the switch.

4   Refitting is a reversal of removal.

20 Brake pedal (manual gearbox) – removal and refitting

This is carried out in a similar manner to that described for the clutch pedal in Chapter 5, Section 5.

21 Brake pedal (automatic transmission) – removal and refitting

1   Remove the parcels tray (Chapter 12).

2   Remove the steering column assembly (Chapter 11).

3   Remove the demister duct on the driver's side.

4   Disconnect the accelerator cable from the pedal arm after pulling off the spring clip.

5   Remove the accelerator pedal bracket from the studs on the brake servo unit.

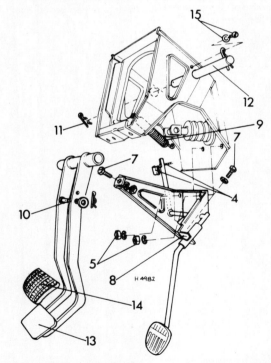

Fig. 9.22 Control pedal arrangement (automatic transmission)
(Sec 21)

4   Accelerator cable and   10   Clevis pin
     spring clip   11   Spring pin
5   Accelerator pedal bracket   12   Pedal shaft
     to servo stud nuts   13   Pedal assembly
7   Setscrew   14   Pedal rubber
8   Accelerator pedal bracket   15   Self-tapping screw
9   Brake pedal return spring

6 Pull the carpet down from behind the panels and remove the insulation pad from the outside top surface of the engine compartment rear bulkhead.

7 Remove the setscrew which secures the accelerator pedal bracket to the body and pedal bracket.

8 Withdraw the accelerator pedal bracket assembly.

9 Disconnect the brake pedal return spring.

10 Remove the clevis pin which secures the pushrod to the brake pedal.

11 Remove the self-tapping screws and washers which secure the pedal shaft spring clips to the pedal bracket. Withdraw the spring clips.

12 Push the shaft through the pedal and withdraw the pedal.

13 Renewal of the pedal shaft bushes can be carried out by pressing the old ones out and the new ones in until they are just below the end face of the tube.

14 Refitting is a reversal of removal but apply grease to all friction surfaces.

22 Pedal bracket – removal and refitting

1 Remove the parcels tray (see Chapter 12).

2 *On manual gearbox cars,* disconnect the steering column and lower it but leaving it still attached to the intermediate shaft (see Chapter 11).

3 *On automatic transmission cars,* the steering column must be removed from the car.

4 Release the strap which secures the cables to the pedal bracket.

5 Disconnect the speedometer cable from the back of the speedometer and withdraw the cable through the pedal bracket.

6 Disconnect the cables from the brake stoplamp switch.

7 Disconnect the choke control cable from the carburettor and withdraw the cable through the pedal bracket hole.

8 If a radio is fitted, pull the aerial (antenna) lead from the back of the radio and withdraw it through the pedal bracket.

9 Carry out the operations described in paragraphs 4 to 8 and 10 of the preceding Section.

10 *On manual gearbox cars,* remove the clevis pin which connects the pushrod to the clutch pedal, also remove the nuts which secure the clutch master cylinder.

11 *On automatic transmission cars,* remove the nuts which secure the backing plate.

12 Remove the two screws which secure the top of the pedal bracket to the body and withdraw the pedal bracket assembly.

13 Remove the brake stoplamp switch and detach the brake pedal return spring.

14 Disconnect the pedals from the bracket, as described in Sections 20 or 21 according to type.

15 Refitting is a reversal of removal but make sure that the accelerator pedal return spring is hard against the pedal lever and under the panel.

23 Brake stoplamp switch – adjustment

1 This is easier to adjust if the parcels tray is first removed (see Chapter 12).

2 Switch on the ignition and release the switch locknuts. Adjust the position of the switch by turning the nuts until the stoplamps go out, then turn the nuts a further half turn to move the switch closer to the pedal arm and lock the nuts. Switch off the ignition.

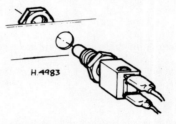

H.4983

Fig. 9.23 Brake stop lamp switch (Sec 23)

24 Fault diagnosis — braking system

Before diagnosing faults from the following chart, check that any braking irregularities are not caused by:
  Uneven and incorrect tyre pressures
  Wear in the steering mechanism
  Defects in the suspension and dampers
  Misalignment of the bodyframe

Symptom	Reason(s)
Pedal travels a long way before the brakes operate	Incorrect pedal adjustment Brake shoes set too far from the drums (seized adjuster)
Stopping ability poor, even though pedal pressure is firm	Linings, discs or drums badly worn or scored One or more wheel hydraulic cylinders seized, resulting in some brake shoes not pressing against the drums (or pads against disc) Brake linings contaminated with oil Wrong type of linings fitted (too hard) Brake shoes wrongly assembled Servo unit not functioning
Car veers to one side when the brakes are applied	Brake pads or linings on one side are contaminated with oil Hydraulic wheel cylinder on one side partially or fully seized A mixture of lining materials fitted between sides Brake discs not matched Unequal wear between sides caused by partially seized wheel cylinders
Pedal feels spongy when the brakes are applied	Air is present in the hydraulic system
Pedal feels springy when the brakes are applied	Brake linings not bedded into the drums (after fitting new ones) Master cylinder or brake backplate mounting bolts loose Severe wear in brake drums causing distortion when brakes are applied Discs out of true
Pedal travels right down with little or no resistance and brakes are virtually non-operative	Leak in hydraulic system resulting in lack of pressure for operating wheel cylinders If no signs of leakage are apparent the master cylinder internal seals are failing to sustain pressure
Binding, juddering, overheating	One or a combination of reasons given above Shoes installed incorrectly with reference to leading and trailing ends Broken shoe return spring Disc distorted Drum distorted
Lack of servo assistance	Vacuum hose disconnected or leaking Non-return valve defective or incorrectly fitted Servo internal defect Clogged servo air filter

Chapter 10 Electrical system

Contents

Specifications

System ....

12 volt negative earth, battery, alternator and pre-engaged starter motor

Battery ....

Sealed for life type

Alternator

Type ....

Lucas A133 65A

Regulated voltage ....

13.6 to 14.4 volt

Brush length:

New ....

0.788 in (20.0 mm)

Minimum projection from holder ....

0.394 in (10.0 mm)

Starter motor

Type ....

Lucas 2M 100 pre-engaged

Minimum brush length ....

0.394 in (10.0 mm)

Armature endfloat ....

0.01 in (0.25 mm)

Fuses

Fuse number	Circuit protected	Amps
1A	Heated rear screen	25
1B	Left-hand parking, tail and number plate lamps.	
	Front foglamp relay	10
1C	Radio, clock	3
2A	Front electric windows	30
2B	Right-hand parking, tail and number plate lamps.	
	Window lift switch illumination. Instrument, heater	
	control and switch illumination	5
2C	Heater blower motor, window lift relays	15
3A	Spare	30
3B	Left-hand headlamp (dipped)	10
3C	Tailgate wiper/washer	10
4A	Clock	3
4B	Right-hand headlamp (dipped)	10
4C	Front foglamps	15
5A	Hazard warning, cigar lighter	15
5B	Left-hand headlamp (main beam)	10
5C	Direction indicators. Heated rear screen relay.	
	Reverse and stop lamps	10
6A	Luggage compartment and interior lamp	5
6B	Right-hand headlamp (main beam)	10
6C	Windscreen wiper/washer	15
7A	Rear fog and warning lamps	5
7B	Bulb failure indicator	3
7C	Door lock relays	5

Bulbs

	Wattage
Headlamp	60/65
Front parking lamp	4
Direction indicator lamps	21
Stop/tail lamp	21/5
Reversing, rear foglamps	21
Number plate lamp	4
Luggage compartment lamp	6
Interior lamp	6
Glovebox lamp	6
Cigar lighter lamp	2.2
Selector lever lamp (auto. trans)	2
Warning lamps, switch illumination	2
Heater control lamp	1.2

Torque wrench settings

	lbf ft	Nm
Alternator adjuster link bolt	12	17
Alternator mounting bolts	20	28
Alternator pulley	25	35
Starter motor securing bolts	30	41

1 Description and maintenance

1   The electrical system is of 12 volt negative earth type. The battery supplies a steady current to the ignition system and for the operation of the electrical accessories.
2   The alternator maintains the charge in the battery. The voltage regulator, which is incorporated in the alternator, adjusts the charging rate according to the demands of the engine.
3   A pre-engaged type starter motor is fitted to all models.
4   Maintenance consists of regularly checking all electrical connections for security and keeping the alternator drivebelt correctly tensioned (see Chapter 2, Section 10).
5   Occasionally wipe dirt and grease from the external surfaces of the alternator and the starter motor.

2 Battery – maintenance and inspection

1   The battery fitted as original equipment is of sealed type and requires no maintenance other than keeping the terminals clean and tight. Smear the terminals with petroleum jelly to prevent corrosion.
2   If a replacement non-sealed type battery is fitted then observe the following procedure.
3   Keep the top of the battery clean by wiping away dirt and moisture.

4   Remove the plugs or lid from the cells and check that the electrolyte level is just above the separator plates. If the level has fallen, add only distilled water until the electrolyte level is just above the separator plates.
5   As well as keeping the terminals clean and covered with petroleum jelly, the top of the battery, and especially the top of the cells, should be kept clean and dry. This helps prevent corrosion and ensures that the battery does not become partially discharged by leakage through dampness and dirt.
6   Once every six months, remove the battery and inspect the battery securing bolts, the battery clamp plate, tray and battery leads for corrosion (white fluffy deposits on the metal which are brittle to touch). If any corrosion is found, clean off the deposits with ammonia and paint over the clean metal with an anti-rust/anti-acid paint.
7   If topping up of the battery becomes excessive and the case has been inspected for cracks that could cause leakage, but none are found, the battery is being overcharged and the voltage regulator within the alternator must be at fault.
8   The battery electrolyte can be measured for specific gravity using a hydrometer. There should be very little variation between the cells and if a variation in excess of 0.25 is present, it will be due to:

(a)   *Loss of electrolyte from the battery at some time caused by spillage or a leak, resulting in a drop in the specific gravity of electrolyte when the deficiency was replaced with distilled water instead of fresh electrolyte. Always leave the addition of acid to a battery to your garage or service station*

(b)   An internal short circuit caused by buckling of the plates of a similar malady pointing to the likelihood of total battery failure in the near future

9   The specific gravity of the electrolyte for fully charged conditions at the electrolyte temperature indicated, is listed in Table A. The specific gravity of a fully discharged battery at different temperatures of the electrolyte is given in Table B.

Table A

Specific Gravity – Battery Fully Charged

1.268 at 100°F or 38°C electrolyte temperature
1.272 at 90°F or 32°C electrolyte temperature
1.276 at 80°F or 27°C electrolyte temperature
1.280 at 70°F or 21°C electrolyte temperature
1.284 at 60°F or 16°C electrolyte temperature
1.288 at 50°F or 10°C electrolyte temperature
1.292 at 40°F or 4°C electrolyte temperature
1.296 at 30°F or -1.1°C electrolyte temperature

Table B

Specific Gravity – Battery Fully Discharged

1.098 at 100°F or 38°C electrolyte temperature
1.102 at 90°F or 32°C electrolyte temperature
1.106 at 80°F or 27°C electrolyte temperature
1.110 at 70°F or 21°C electrolyte temperature
1.114 at 60°F or 16°C electrolyte temperature
1.118 at 50°F or 10°C electrolyte temperature
1.122 at 40°F or 4°C electrolyte temperature
1.126 at 30°F or -1.1°C electrolyte temperature

3   Battery – charging

1   In winter time when heavy demand is placed upon the battery, such as when starting from cold, and much electrical equipment is continually in use, it may be necessary to charge the battery from an outside source particularly when the car is used only for short journeys.
2   A trickle charger charging at 1.5A can be safely used overnight.
3   Specially rapid 'boost' charges which are claimed to restore the power of the battery in 1 to 2 hours are most dangerous as they can cause serious damage to the battery plates, particularly in maintenance-free type batteries.

4   Battery – removal and refitting

1   The battery is located within the engine compartment on the right-hand side (photo).
2   To remove the battery first disconnect the lead from the negative (-) terminal followed by the positive (+) terminal.
3   Detach the battery holding down clamp bolts and lift the battery from its platform. Take care not to spill electrolyte on the bodywork or the paint will be damaged.

4   Refitting is a reversal of removal but make sure that the lead terminals and battery posts are clean and making a sound metal-to-metal contact.

5   Alternator – description and precautions

1   The alternator is of Lucas 65A type with integral voltage regulator.
2   Take extreme care when making circuit connections to a vehicle fitted with an alternator and observe the following. When making connections to the alternator from a battery always match correct polarity. Before using electric-arc welding equipment to repair any part of the vehicle, disconnect the connector from the alternator and disconnect the positive battery terminal. Never start the car with a battery charger connected. Always disconnect both battery leads before using a main charger. If boosting from another battery, always connect in parallel using heavy cable.
3   Never pull off a battery lead while the engine is running as a means of stopping it. If working under the bonnet and the engine must be stopped, pull the coil high tension lead or low tension lead from the distributor.

6   Alternator – removal, overhaul and refitting

1   Disconnect the battery.
2   From the rear of the alternator, release the clip and disconnect the multi-plug.
3   Release the alternator mounting and adjuster link bolts, push the alternator in towards the engine and slip off the drivebelt.
4   Unscrew the mounting and adjuster link bolts and remove the alternator (photo).
5   Before overhauling an alternator, if the unit has been in service for a long time, it may be found more economical to exchange it for a factory reconditioned unit.
6   If overhaul is to be carried out, first clean away external dirt and remove the radio interference capacitor.
7   Extract the screws and remove the cover.
8   Remove the surge protection diode and the brushbox (photo).
9   Using a soldering iron disconnect the rectifier tags. Prise out the cable ends and then remove the assembly from the end bracket.
10   Unscrew and remove the tie-bolts and pull off the slip ring end bracket and stator. Extract the O-ring seal from inside the slip ring end bracket.
11   Unscrew the pulley nut. On some alternators, a recess is provided in the end of the shaft so that an Allen key can be inserted to hold the shaft still while the nut is unscrewed. If such a recess is not provided, place an old drivebelt in the pulley groove and then grip the ends of the belt as close to the pulley as possible in the jaws of a vice.
12   Remove the pulley and fan.
13   Prise out the Woodruff key and then press the rotor shaft out of the drive end bracket bearing.
14   Extract the circlip to release the bearing, bearing cover plate, O-ring and felt washer from the bracket.
15   Inspect all components for wear.
16   Clean the slip rings with a fuel moistened rag. If necessary they may be cleaned with very fine glass paper not sand paper (photo).

4.1 Battery and fixing clamp

6.4 Alternator top mounting

6.7A Alternator rear cover

6.7B Removing alternator rear cover

6.8 Alternator, rear cover removed

6.16 Alternator slip rings

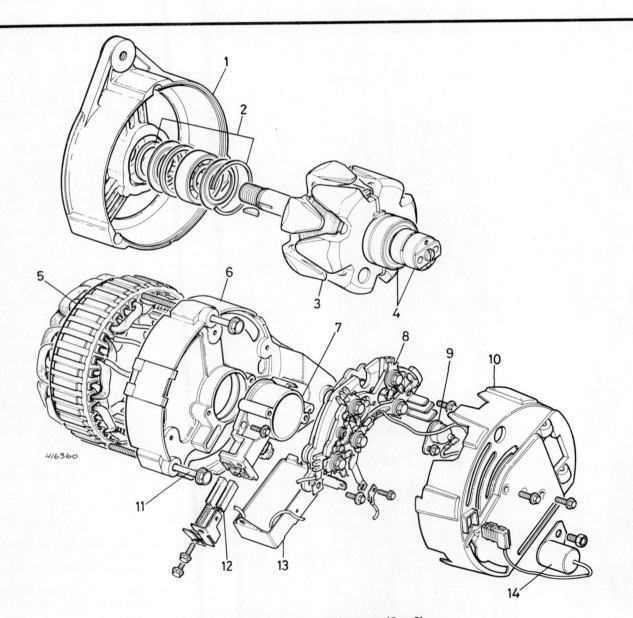

Fig. 10.1 Exploded view of alternator (Sec 6)

1	Front end bracket	5	Stator	9	Surge protection diode
2	Bearing assembly	6	Rear end bracket	10	Cover
3	Rotor assembly	7	Brush box	11	Fixing bolts
4	Slip ring end bearing	8	Rectifier pack		

12	Brushes
13	Regulator
14	Radio interference capacitor

17 Electrical checks of the slip rings, field windings and stator and diodes are not within the scope of the home mechanic owing to the need for special test equipment.

18 Check that the brush spring pressure is satisfactory when the brushes are depressed into the brush box (photo).

19 Extract the screw which holds the voltage regulator to the brush box.

20 Note the colour and location of the leads and terminal strips and remove the brushes. Check the brushes for wear and renew them if outside the specified length.

21 If the rectifier must be renewed, make sure that its replacement is of exactly similar type.

22 Reassemble by reversing the dismantling operations, but observe the following points.

23 Apply Shell Alvania RA grease to the rotor shaft bearings.

24 Tighten the pulley nut to the specified torque.

25 When pressing the rotor into the drive-end bracket, support the bearing inner track.

26 Before refitting the alternator, check that the bracket mounting bolts are tight (photo).

27 Refit the alternator, tension the drivebelt as described in Chapter 2, Section 10.

28 Reconnect the electrical plug and clip to the rear face of the alternator.

6.18 Checking brush spring pressure

6.26 Alternator mounting bracket

7 Starter motor – description and testing in car

1 The method of engagement on the pre-engaged starter is that the drive pinion is brought into mesh with the starter ring gear before the main starter current is applied.

2 When the ignition is switched on, current flows from the battery to the solenoid which is mounted on the top of the starter motor body. The plunger in the solenoid moves inward so causing a centrally pivoted lever to move in such a manner that the forked end pushes the drive pinion into mesh with the starter ring gear. When the solenoid plunger reaches the end of its travel, it closes an internal contact and full starting current flows to the stator field coils. The armature is then able to rotate the crankshaft so starting the engine.

3 A special one way clutch is fitted to the starter drive pinion so that when the engine just fires and starts to operate on its own, it does not drive the stator motor.

4 This type of starter motor causes very little wear to the flywheel (or driveplate) ring gear teeth.

5 If the starter motor fails to turn the engine when the switch is operated there are four possible causes:

(a) The battery is faulty
(b) The electrical connections between the switch, solenoid, battery and starter motor are somewhere failing to pass the necessary current from the battery through the starter to earth
(c) The solenoid switch is faulty
(d) The starter motor is mechanically or electrically defective

6 To check the battery, switch on the headlights. If they dim after a few seconds the battery is in a discharged state. If the lights glow brightly, operate the starter switch and see what happens to the lights. If they dim then you know that power is reaching the starter motor but failing to turn it. If the starter turns slowly when switched on, proceed to the next check.

7 If, when the starter switch is operated the lights stay bright, then insufficient power is reaching the motor. Remove the battery connections, starter/solenoid power connections and the engine earth strap and thoroughly clean them and refit them. Smear petroleum jelly around the battery connections to prevent corrosion. Corroded connections are the most frequent causes of electric system malfunctions.

8 When the above checks and cleaning tasks have been carried out but without success, you will probably have heard a clicking noise each time the starter switch was operated. This was the solenoid switch operating, but it does not necessarily follow that the main contacts were closing properly (if no clicking has been heard from the solenoid, it is certainly defective). The solenoid contact can be checked by putting a voltmeter or bulb across the main cable connection on the starter side of the solenoid and earth. When the switch is operated, there should be a reading or lighted bulb. If there is no reading or lighted bulb, the solenoid unit is faulty and should be renewed.

9 Finally, if it is established that the solenoid is not faulty and 12 volts are getting to the starter, then the motor is faulty and should be removed for inspection.

8 Starter motor – removal and refitting

1 Open the bonnet and disconnect the battery.

2 Peel back the rubber cover from the terminal on the starter solenoid.

3 Unscrew the terminal nut and disconnect the cable (photo).

4 Pull the remaining lead from its terminal.

5 Unscrew the starter motor rear support bracket from the cylinder block (photo).

6 Unscrew and remove the two starter motor flange mounting bolts, support the unit with the hand and withdraw it. A spacer plate is fitted to some units (photos).

7 Refitting is a reversal of removal.

9 Starter motor solenoid – removal and refitting

1 The starter motor solenoid may be removed while the unit is still in the car.

8.3 Starter solenoid connections

8.5 Starter motor rear mounting bracket

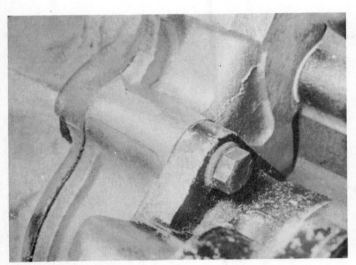

8.6A Starter motor mounting bolt

8.6B Removing starter motor

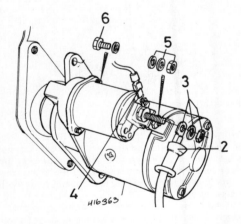

Fig. 10.2 Starter motor solenoid connections (Sec 8)

2 Rubber cover   5 Connecting link nut and washer
3 Terminal nut and washer   6 Solenoid set screws
4 Lucar type connector   7 Solenoid

2   Open the bonnet and disconnect the battery.
3   Peel back the rubber cover from the terminal on the solenoid.
4   Unscrew the terminal nut and disconnect the cable.
5   Disconnect the remaining lead from the solenoid.
6   Unscrew the nut from the connecting link which runs to the starter motor terminal.
7   Extract the solenoid fixing screws and pull the solenoid from the drive-end bracket.
8   Pull the spring and spring seat from the operating plunger and then remove the dust excluder.
9   Depress the operating plunger and unhook it from the engagement lever.
10   Refitting is a reversal of removal, but smear the plunger link and engagement lever with Shell Retinax A grease.

10 Starter motor – overhaul

1   With the starter removed from the car, clean away external dirt.
2   Slacken the nut which secures the connecting link to the solenoid terminal 'STA'.
3   Remove the two screws which secure the solenoid to the drive-end bracket.

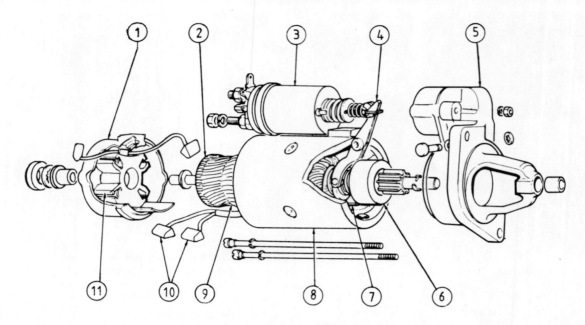

Fig. 10.3 Exploded view of starter motor (Sec 10)

1	Commutator end bracket	4	Engagement lever	7
2	Commutator	5	Drive end bracket	8
3	Solenoid	6	Drive assembly	9

1 Commutator end bracket
2 Commutator
3 Solenoid
4 Engagement lever
5 Drive end bracket
6 Drive assembly
7 Pole shoe
8 Yoke
9 Armature
10 Brush
11 Brush box moulding

10.6 Starter motor armature end cap

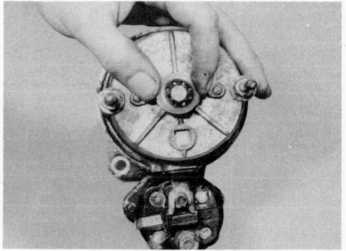

10.7 Armature shaft spire nut

4 Lift the solenoid plunger upwards and separate it from the engagement lever. Extract the return spring, spring seat and dust excluder from the plunger body.
5 Withdraw the block from between the drive end bracket and the starter motor yoke.
6 Remove the armature end cap from the commutator end bracket (photo).
7 Chisel off some of the claws from the armature shaft spire nut so that the nut can be withdrawn from the shaft (photo).
8 Remove the two tie-bolts and the starter motor support bracket (photo).
9 Withdraw the starter motor end cover and yoke from the drive end bracket.

10 Remove the two tie-bolts and then withdraw the commutator end cover and starter motor yoke from the drive-end bracket (photo).
11 Separate the commutator end cover from the starter motor yoke, at the same time disengaging the field coil brushes from the brush box to facilitate separation.
12 Withdraw the thrust washer from the armature shaft.
13 Remove the spire nut from the engagement lever pivot pin and then extract the pin from the drive-end bracket.
14 Withdraw the armature and roller clutch drive assembly from the drive-end bracket.
15 Using a piece of tubing, drive back the thrust collar to expose the jump ring on the armature shaft. Remove the jump ring and withdraw the thrust collar and the roller clutch.

10.8 Starter motor tie-bolts and bracket

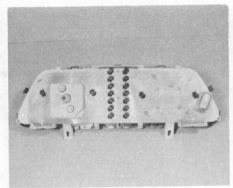

10.10 Rear view of instrument panel

10.19 Soldering starter motor brush leads

16 Remove the spring ring and release the engagement lever, thrust washers and spring from the roller-clutch drive.

17 Remove the dust-excluding seal from the bore of the drive-end bracket.

18 Inspect all components for wear. If the armature shaft bushes require renewal, press them out or screw in a $\frac{1}{2}$ in tap. Before inserting the new bushes, soak them in engine oil for 24 hours.

19 If the brushes have worn below the minimum specified length, renew them by cutting the end bracket brush leads from the terminal post. File a groove in the head of the terminal post and solder the new brush leads into the groove. Cut the field winding brush leads about $\frac{1}{4}$ in (6.4 mm) from the joint of the field windings. Solder the new brush leads to the ends of the old ones. Localise the heat from the field windings (photo).

20 Check the field windings for continuity using a 12V battery and test bulb. If the windings are faulty, removal of the pole shoe screws should be left to a service station having a pressure screwdriver as they are very tight.

21 Discoloration of the commutator should be removed by polishing it with a piece of glass paper (**not** emery cloth). Do not undercut the insulation.

22 Reassembly is a reversal of dismantling but apply grease to the moving parts of the engagement lever, the outer surface of the roller clutch housing and to the lips of the drive-end bracket dust seal. Fit a new spire to the armature shaft, positioning it to give the specified shaft endfloat. Measure this endfloat by inserting feeler blades between the face of the spire nut and the flange of the commutator end bush.

11 Fuses and relays

1 The fusebox is located under the facia panel (photo).

2 Access to the fuses is obtained by turning the quick release buttons and pulling off the panel cover (photo).

3 Spare fuses and a diagram to indicate which circuits the fuses protect are located behind the panel cover (photo).

4 Always renew a blown fuse with one of identical rating.

5 An in-line type frame is incorporated in the radio feed wire.

6 A number of relays is fitted and some or all of the following may be fitted according to the specification of the particular model. The relays are located either on the fuse block or within the engine compartment.

    1   Fuel pump relay
    2   Rear window lift motor relay
    3   Front window lift motor relay
    4   Starter solenoid relay
    5   Heated rear screen relay
    6   Front foglamp relay

7 The direction indicator flasher relay is also mounted on the fuse block.

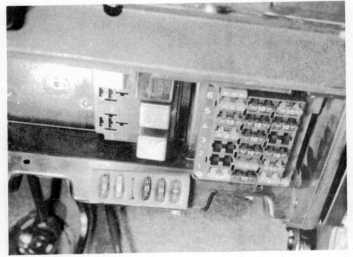

11.1 Fuse box

11.2 Fuse box lid

11.3 Spare fuses

12.4 Steering column combination switch

12 Steering column switch – removal and refitting

1   Disconnect the lead from the battery negative terminal.
2   Remove the steering wheel, as described in Chapter 11.
3   Release the strap which secures the switch wiring harness to the steering column and then disconnect the multi-pin connectors.
4   Slacken the switch clamp screw and withdraw the switch assembly from the steering column (photo).
5   The wiper/washer switch can be detached from the mounting plate of Lucas type switches if the two securing rivets are drilled out.
6   When refitting the switch, check that the cancelling ring is free to rotate in the switch. Locate the lug on the inner diameter of the switch in the slot in the steering column tube.
7   Align the arrow on the switch so that with right-hand steering it is towards the horn stalk or with left-hand steering towards the pip on the cover.
8   Once the switch is fitted, check that the cancelling ring is free to rotate. If it is tight, slacken the screws and reposition the top plate.
9   Reconnect the wiring plugs, the battery and refit the steering wheel.

13 Control switches – removal and refitting

1   Before any switch is removed, disconnect the battery negative lead.

Lighting switch
2   This switch is located in the left-hand steering column shroud.
3   Remove the shroud, disconnect the wiring plug.
4   Depress the switch securing tabs and remove the switch.

Luggage compartment switch
5   Raise the tailgate and remove the trim pad.
6   Release the mercury type switch from its clip and disconnect the leads.

Facia push-type switches
7   Pull out the finisher plate located between the two rows of switches.
8   Using a hooked piece of wire, inserted through the centre hole in the switch pack, pull the switch assembly from its housing.
9   The individual switches may be removed by squeezing the retaining lugs together.

Courtesy lamp switch
10 These switches are located in the front door pillars (photo).

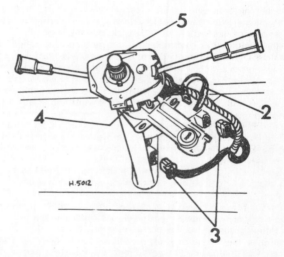

Fig. 10.4 Steering column combination switch (Sec 12)

2   Harness strap   4   Switch clamp screw
3   Multi-pin connector   5   Switch

Fig. 10.5 Lighting switch (Sec 13)

11 Extract the fixing screw and withdraw the switch.
12 If the leads are to be disconnected, tape them to the pillar in case they slip inside the pillar cavity.
13 The switch should be smeared with petroleum jelly before refitting in order to prevent corrosion.

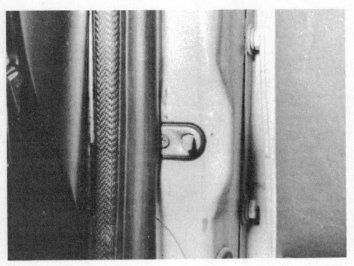
13.10 Courtesy lamp switch

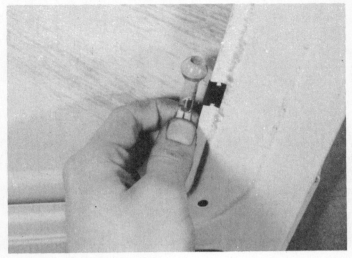
13.14 Tailgate magnetic switch

Tailgate switch
14 This is a mercury type switch located behind a small blanking panel in the lower part of the tailgate lid (photo).
15 Take great care when handling it as it is very fragile.
16 Refitting of all control switches is a reversal of removal.

14 Cigar lighters – removal and refitting

1 Disconnect the battery.
2 Remove the glovebox
3 Pull out the heater element from the cigar lighter.
4 Reach up behind the facia panel and unscrew the cigar lighter rear shell from its base. Complete the unscrewing from the front of the facia.
5 Disconnect the wiring and remove the lighter.

15 Headlamp bulb – renewal

1 Open the bonnet and support it on its stay.
2 Twist the protective cover on the back of the headlamp in an anticlockwise direction and pull it away from the lamp (photo).

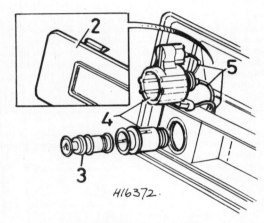
H16372.

Fig. 10.6 Cigar lighter (Sec 14)

2 Glovebox lid   4 Shell
3 Heating element   5 Electrical leads

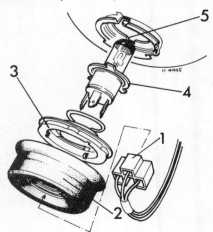
Fig. 10.7 Headlamp bulbholder (Sec 15)

1 Connector plug   4 Bulb
2 Dust excluder   5 Bulbholder alignment notch
3 Retainer

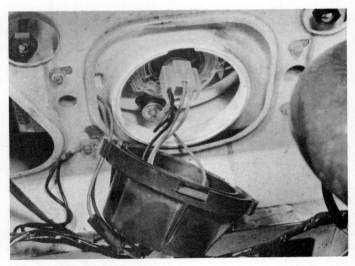
15.2 Headlamp protective cover

15.3A Headlamp bulbholder springs

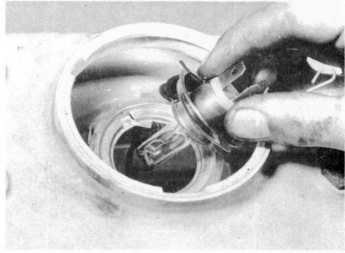

15.3B Headlamp bulbholder

3   Pull off the three-pin connector and then release the bulb holder clips and remove the bulb and holder (photos).
4   Do not touch the glass of the new bulb with the fingers, but if this does happen inadvertently, wipe the glass clean with methylated spirit.
5   Fit the bulb making sure that the bulbholder lugs engage correctly in their cut-outs in the reflector.

16   Headlamp beam – alignment

1   Headlamp beam alignment is best left to a service station having optical beam setting equipment.
2   In an emergency the adjustment screws are accessible from inside the engine compartment.

17   Headlamp – removal and refitting

1   Open the bonnet and prop it with the stay.
2   Remove the protective cover from the rear of the headlamp and disconnect the multi-plug.

3   Remove the parking lamp bulb and holder.
4   Disconnect the remaining feed and earth leads (photo).
5   Unscrew the headlamp fixing nuts and withdraw the headlamp unit complete with direction indicator lamp. This can be detached from the headlamp (photo).
6   Refitting is a reversal of removal. Have the headlamp beams aligned on completion.

18   Exterior lamps – bulb renewal

Front parking lamp
1   Refer to Section 15 to gain access to the bulbholder which is located in the headlamp reflector.
2   Pull the bulbholder from the lamp reflector.
3   Remove the bulb which is of bayonet fitting type.

Front direction indicator lamp
4   Extract the two lens screws, remove the lens (photo).
5   Remove the bulb which is of bayonet fitting type (photo).

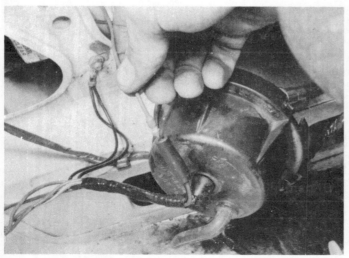

17.4 Headlamp earth lead

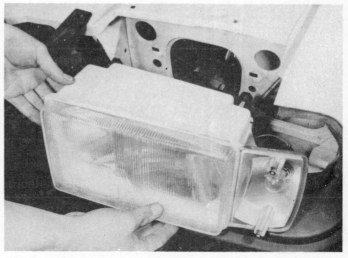

17.5 Removing headlamp

Rear lamp cluster

6   Extract the four screws and remove the lens (photo).
7   Remove the stop, tail, reversing lamp or rear foglamp bulb as necessary, all bulbs being of bayonet fitting type (photo).

Rear direction indicator lamp

8   Remove the lens from the rear lamp cluster.
9   Extract the two screws and remove the direction indicator lamp lens (photo).
10   Remove the bayonet fitting type bulb.

Rear number plate lamp

11   Prise the lamp from the rear bumper, then pull the bulbholder from the lamp (photos).
12   Remove the bayonet fitting type bulb (photo).

Front and rear foglamps

13   The bayonet fitting type bulbs are accessible after removing the lamp lenses.
14   Fitting the new bulb to all lamps is a reversal of the removal operations.

19 Interior lamps — bulb renewal

Courtesy (interior) lamp

1   Pull or carefully prise the lamp from the roof lining (photo).
2   Pull the festoon type bulb from its contacts.

Glovebox lamp

3   Open the glovebox and prise the bulb from its contacts. Use a thin screwdriver as a lever under the cap of the bulb.

18.4 Extracting front direction indicator lamp lens

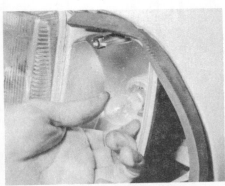
18.5 Front direction indicator bulb

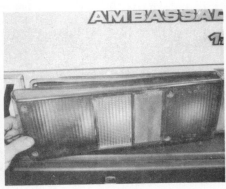
18.6 Rear lamp cluster lens

18.7 Rear lamp bulb

18.9 Rear direction indicator lamp lens screw

18.11A Rear number plate lamp

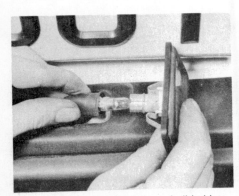
18.11B Rear number plate lamp bulbholder

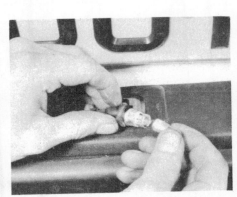
18.12 Rear number plate lamp bulb

19.1 Interior lamp

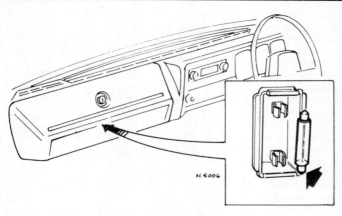

Fig. 10.8 Glovebox lamp (Sec 19)

19.14 Instrument panel bulb and holder

Luggage compartment lamp
4   Prise the lamp from its bracket and pull the festoon type bulb from its contacts.

Heater control panel bulb
5   Remove the radio (Section 31)
6   Remove the parcel tray cover and the right-hand hot air duct from the heater.
7   Withdraw the bulbholder from the heater control panel.
8   Pull the wedge base type bulb from its holder.

Cigar lighter bulb
9   Remove the glovebox.
10   Remove the cigar lighter as described in Section 14.
11   Squeeze the sides of the bulb hood to release it.
12   Remove the small lens and the bulb holder and then withdraw the bayonet fitting type bulb.

Instrument panel and warning lamp bulbs
13   Remove the cover panels from under the facia.
14   Reach up behind the instrument panel and turn the bulbholder with the blown bulb anti-clockwise and then pull it from the panel (photo).
15   The wedge base type bulb can be pulled from the holder.

Automatic transmission selector index bulb
16   Prise the flexible gaiters up the seat belt stalks.
17   Slacken the locknut and unscrew and remove the selector lever knob.
18   Extract the centre console fixing screws.
19   Set the selector lever in 1 or 2 and then prise up the quadrant.
20   The bayonet fitting type bulbs may now be removed from their holders.
21   Refitting of all the interior bulbs is a reversal of removal.

20 Speedometer cable – renewal

1   At the transmission end of the cable, unscrew the knurled nut that holds the cable to the pinion assembly.
2   Working inside the car, remove the cover panel from under the facia panel below the instrument panel.
3   Reach up behind the instrument panel and squeeze the knurled sections of the cable connector together and pull the cable from the speedometer (photo).
4   Prise out the cable grommet and pull the cable assembly through the bulkhead into the engine compartment.
5   The inner and outer cables are supplied as an assembly. Refitting is a reversal of removal. Do not route the cable so that bends are of smaller radius than the original installation.
6   Note that the angled drive shown in Fig. 10.10 for automatic transmission models was only fitted between VIN 100124 and 108201. Other automatic transmission models have a similar arrangement to manual cars.

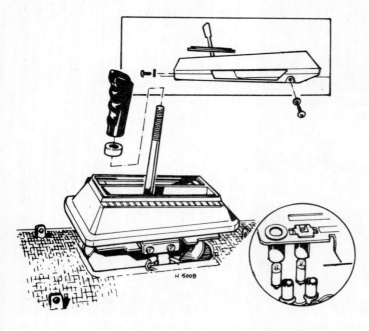

Fig. 10.9 Automatic transmission speed selector illumination (Sec 19)

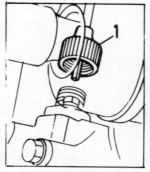

Fig. 10.10 Speedometer cable connection at transmission (Sec 20)

1   *Manual transmission*   2   *Automatic transmission*

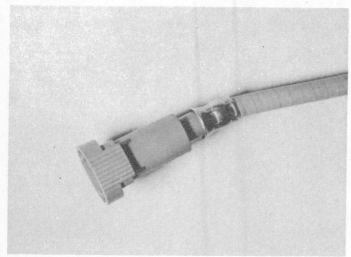
20.3 Speedometer cable connector

21 Instrument panel – removal and refitting

1 Disconnect the battery.
2 Remove the steering wheel as described in Chapter 11.
3 Extract the three screws from under the top lip of the instrument panel hood (photo).
4 Carefully prise up the lower lip of the hood and withdraw the hood (photo).
5 Remove the right-hand switch panel. This is secured by one screw at the top and two screws under its lower edge. As the panel is withdrawn, disconnect the wiring plugs (photos).
6 Remove the left-hand cubby hole. This is held by one screw at the top and two screws under its lower edge.
7 Remove the two screws which are now accessible at the base of the instrument panel.
8 Lift the top of the panel and tilt it towards you. Reach behind the panel and disconnect the speedometer cable by squeezing the knurled sections of the cable connector together. It may be necessary to have an assistant feed the speedometer cable through the bulkhead grommet to enable the panel to be withdrawn far enough to be able to reach the cable connector (photo).

21.3 Extracting instrument panel hood screw

21.4 Removing instrument panel hood

21.5A Extracting switch panel screw

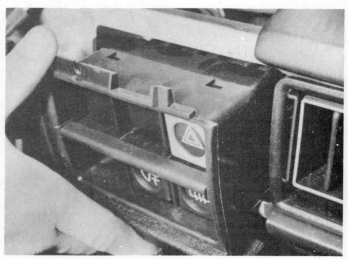
21.5B Withdrawing switch panel

21.5C Wiring multi-plugs

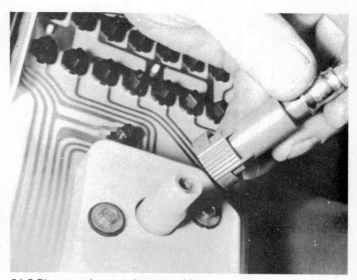

21.8 Disconnecting speedometer cable

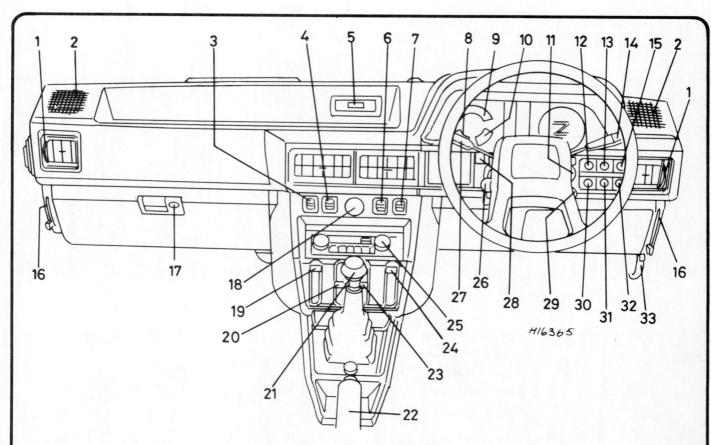

H16365

Fig. 10.11 Facia layout (Sec 21)

1	Side vent control	10	Fuel gauge
2	Radio speaker	11	Panel light dimmer
3	Spare switch position	12	Rear window wiper switch
4	LH door electric window switch	13	Front foglamp switch
5	Digital clock	14	Windscreen wiper washer control
6	RH door electric window switch	15	Hazard warning switch
7	Spare switch position	16	Footwell vent control
8	Temperature gauge	17	Glovebox lock
9	Econometer		

18	Cigar lighter
19	Centre vent
20	Temperature control
21	Gear lever
22	Handbrake
23	Air distribution control
24	Air intake control/blower switch
25	Radio/cassette player

26	Choke control
27	Combination switch
28	Light switch
29	Ignition/starter switch
30	Tailgate window wash switch
31	Rear foglamp switch
32	Rear demister switch
33	Bonnet release handle

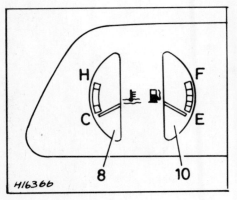 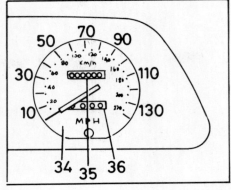 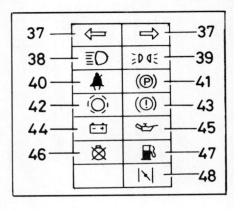

Fig. 10.12 Instrument and warning lamps (Sec 21)

8 Temperature gauge	36 Trip recorder	40 Seat belt	44 Ignition/charge
10 Fuel gauge	37 Direction indicator/hazard	41 Handbrake ON	45 Oil pressure
34 Speedometer	warning	42 Brake pad wear	46 Bulb failure
35 Distance recorder	38 Main beam	43 Low fluid level	47 Low fuel level
	39 Parking lamps		48 Choke ON

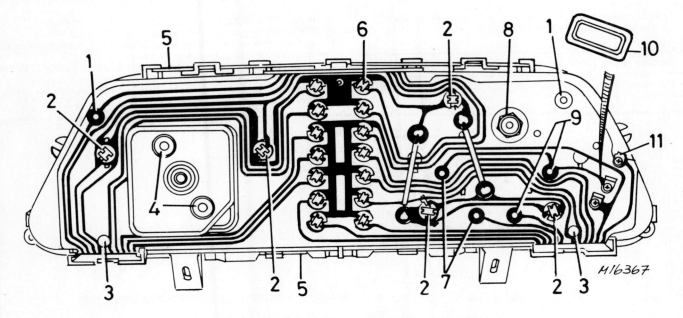

Fig. 10.13 Rear view of instrument panel (Sec 21)

1 Locating peg	4 Speedometer fixing screws	7 Fuel gauge fixing nuts	10 Instrument voltage
2 Panel lamp	5 Retainer	8 Econometer fixing nuts	stabiliser
3 Plastic retainer	6 Warning lamps	9 Temperature gauge fixing nuts	11 Blade connector

9   Disconnect the multi-plugs and the vacuum pipe (econometer if fitted). Remove the panel.
10   Individual instruments or the voltage stabiliser may be removed as necessary. The instruments are held by nuts while the stabiliser is of plug-in type. Take great care not to damage the printed circuit.

22 Windscreen wiper blades and arms – removal and refitting

1   The wiper blades should be renewed as soon as they cease to wipe the glass clean.
2   To renew a blade, pull the arm from the glass, depress the small tab at the connector and pull the blade pivot pin from the arm (photo).
3   The rubber inserts or the complete blades are available as replacements.

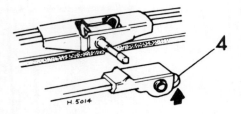

Fig. 10.14 Wiper blade fixing (Sec 22)

4   Clip tab

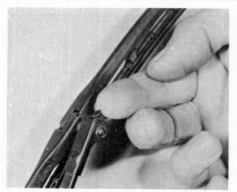

22.2 Releasing wiper blade from arm

22.4 Wiper arm nut

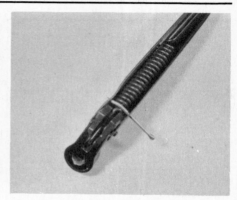

22.6 Temporary wiper arm locking pin

4   To remove a wiper arm, pull the arm from the glass and lock it in this position by passing a pin through the holes in the wiper arm shank. Flip up the plastic cover and unscrew the nut which holds the arm to the drive spindle (photo).

5   Before pulling the arm from the spindle, mark the position of the blade on the glass with a strip of masking tape as an aid to alignment when refitting.

6   Refitting is a reversal of removal, but fit the pin as at removal and do not overtighten the spindle nuts (photo).

23 Tailgate wiper blade and arm – removal and refitting

1   The operations are as described in the preceding Section for the windscreen.

24 Windscreen wiper motor and linkage – removal and refitting

1   Disconnect the battery.
2   Remove the windscreen wiper arm/blade assemblies.
3   Extract the screws from the air intake grille and place it to one side (photo).
4   Release the fixing strap and remove the water shield.
5   Disconnect the wiring harness plug from the wiper motor.
6   Remove the seals and unscrew the drive spindle fixing nuts. Remove the remaining seals and washers.
7   Slide the wiper mounting rubber from its location and release the drive pivots from their holes.
8   Withdraw the motor and the linkage through the air intake aperture (photo).

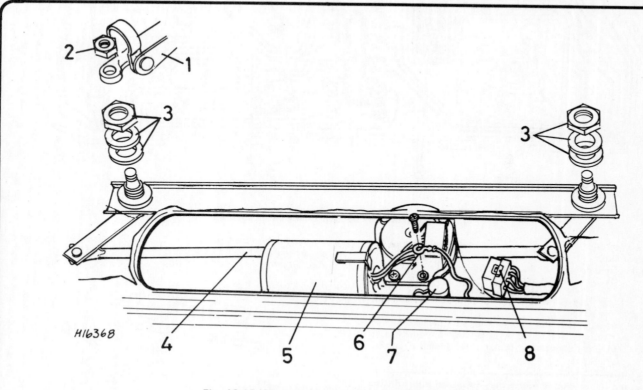

H16368

Fig. 10.15 Windscreen wiper motor and linkage (Sec 24)

1 *Wiper arm*	3 *Drive pivot nut washer and seal*	4 *Link*	6 *Earth lead*
2 *Securing nut*		5 *Wiper motor*	7 *Mounting rubber*
			8 *Multi-pin plug*

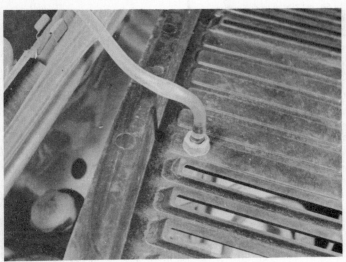

24.3 Windscreen washer pipe and jet at grille

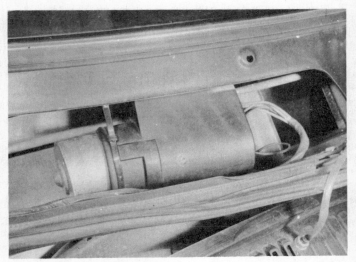

24.8 Windscreen wiper motor

9 Mark the setting of the crank arm on the motor spindle and then unscrew the nut and remove the arm.
10 Unbolt the motor from its frame.
11 Refitting is a reversal of removal.

25 Tailgate wiper motor — removal and refitting

1 Remove the wiper arm/blade assembly and take off the drive spindle nut, washer and seals.
2 Remove the trim panel from the inside of the tailgate and disconnect the electrical leads from the motor.
3 Unbolt the motor mounting bracket and withdraw the motor/gearbox from the tailgate.
4 Refitting is a reversal of removal.

26 Wiper motors — overhaul

1 It is not recommended that the tailgate or windscreen wiper motor is overhauled, but in the event of wear occurring, renew it or have it reconditioned by an auto-electrical engineer.
2 The tailgate wiper motor can be detached in the following way.

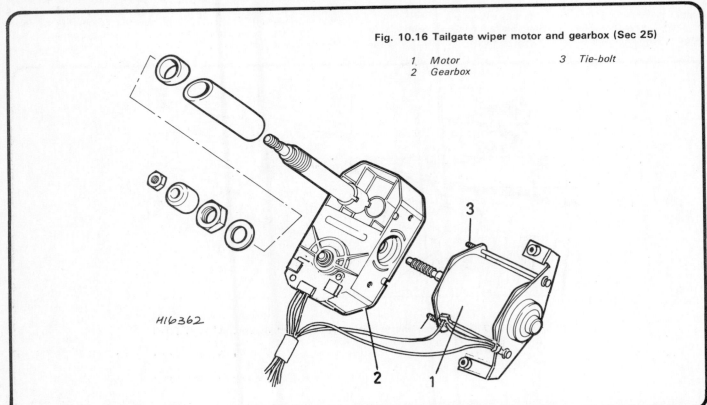

Fig. 10.16 Tailgate wiper motor and gearbox (Sec 25)

1 Motor
2 Gearbox
3 Tie-bolt

H16362

3 Remove the insulating tape from the wiring harness.
4 Unscrew the tie-bolts and separate the motor from the gearbox and the mounting bracket.
5 Fit the new motor to the gearbox and mounting bracket making sure that the earth lead is fitted to the mounting bracket.
6 Screw in the tie-bolts and re-tape the wiring.

27 Screen washer pump and jets

1 The fluid reservoir supplies the windscreen and tailgate glass if the car is equipped with a rear wash/wipe facility.
2 To remove the pump, disconnect the battery, release the fluid reservoir from its mounting bracket and empty out the contents (photo).
3 Disconnect the washer tube and the wiring plug from the pump which is to be removed.
4 Pass a long screwdriver into the reservoir and hold the pump retaining sleeve against rotation by engaging the blade in the sleeve cut-outs.
5 Unscrew and remove the pump.
6 Refitting is a reversal of removal, but use a new sealing washer and do not overtighten the pump into its retaining sleeve.
7 The windscreen or tailgate washer jet pattern can be altered by inserting a pin into the jet nozzle and moving it very slightly in the required direction.
8 Use only screenwash of reputable brand in the washer fluid reservoir. Never use liquid detergent or cooling system anti-freeze as this will damage the pumps and body paintwork.
9 During very cold weather, a little methylated spirit may be added to the reservoir to prevent the fluid from freezing.

28 Heated tailgate window

1 Care should be taken to avoid damage to the element for the tailgate heated window.
2 Avoid scratching with rings on the fingers when cleaning, and do not allow luggage to rub against the side of the glass.
3 Do not stick labels over the element on the inside of the glass.
4 If the element grids do become damaged, a special conductive paint is available from most motor factors to repair it.
5 Do not leave the heated window switched on unnecessarily, as it draws a high current from the electrical system.

27.2 Washer fluid reservoir and pump

29 Power operated electric windows

! This facility is standard equipment on HLS and Vanden Plas models.
2 Operation of the windows is controlled from switches either side of the cigar lighter on the centre console.
3 In the event of a fault occurring, first check the circuit fuse and then the wiring connections for security.
4 Access to the door-mounted electric motors is obtained after removing the door trim panel as described in Chapter 12.
5 Disconnect the battery and then disconnect the wiring plugs from the motor.
6 Withdraw the motor mounting screws and release the window operating arm from the glass channel as described for manually-operated windows in Chapter 12. Withdraw the motor/arm assembly.
7 It is not recommended that the motor should be overhauled, but obtain a new sealed unit.

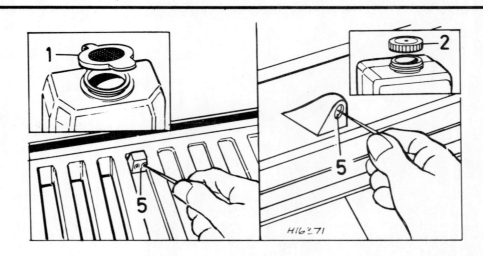

Fig. 10.17 Windscreen and tailgate washer reservoirs and jets (Sec 27)

  1   Reservoir cap   2   Reservoir cap   5   Jet

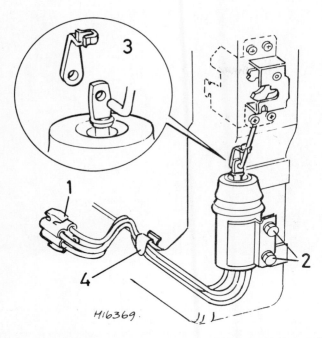

Fig. 10.18 Front door latch solenoid (central locking) (Sec 30)

1 Wiring plug   3 Link
2 Fixing screws   4 Wiring clip

30 Central door lock system

1 With this optional system, the four doors and the tailgate can be locked by operation of the driver's door lock or sill button.
2 Locking may also be carried out without using the key if the exterior lock lever is held in the open position and the sill button depressed and the door then closed in the usual way from outside.
3 Even with the central door locking system in operation, the tailgate lock can be operated independently.
4 The door latch solenoids are accessible once the door trim panel has been removed as described in Chapter 12.
5 Disconnect the latch link from the solenoid (front door) or plastic connector (rear door).
6 Unbolt the solenoid and withdraw it through the door aperture.
7 Refitting is a reversal of removal, but adjust in the following way before refitting the trim panel.

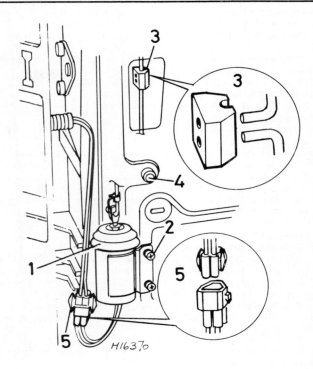

Fig. 10.19 Rear door latch solenoid (central locking) (Sec 30)

1 Solenoid unit   4 Clip
2 Fixing screws   5 Wiring connector
3 Link connector

8 Close the door, depress the sill button and then raise and support the solenoid to eliminate all endplay. Tighten the solenoid screws. Re-check operation, electrically, by using the driver's sill button.

31 Radio – removal and refitting

1 Pull off the control knobs (photo).
2 Unscrew the spindle nuts and remove the finisher and masking plate.
3 Unscrew the radio mounting strips and withdraw the radio until the aerial, power feed and speaker leads can be disconnected. Remove the radio (photo).

31.1 Radio control knobs removed

31.3 Removing radio

4   Refitting is a reversal of removal, but if a new receiver or aerial has been fitted, adjust the trim screw in the following way.
5   Position the car on open ground away from buildings or overhead cables.
6   Fully extend the aerial and tune in to a weak station near 300 metres, medium wave band.
7   Remove the knob and bezel from the tuning control. This will expose the small screw used to trim the aerial.
8   Turn the screw very slowly in either direction to obtain maximum volume.
9   Refit the knob and bezel.

32 Horns

1   The horns are mounted in front of the radiator (photo).
2   No attention to them is required except to keep the wiring and terminals secure.

33 Seat belt warning switch

1   A warning lamp will come on if the ignition is switched on and either front seat is occupied without the seat belt having been fastened.
2   Should a fault occur in the system, first check the security of the wiring at the switch. If this is satisfactory, then the switch must be removed and renewed.
3   Remove the front seat (Chapter 12).
4   Turn the seat upside down and remove the tape which holds the wiring.
5   Unclip the rear edge of the seat trim from the seat frame.
6   Extract the two screws which hold the trim to the seat frame.

32.1 Horn location

7   Remove the clips from the seat cushion overlap trim and then fold back the trim to expose the rubber seat diaphragm.
8   Remove the clips which secure the rubber diaphragm, lift the diaphragm and remove the switch.
9   Refit by reversing the removal operations.

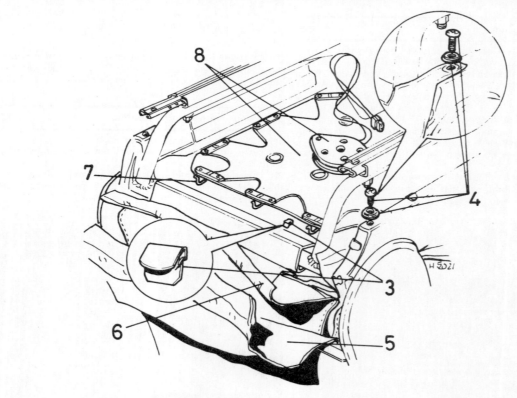

Fig. 10.20 Seat belt switch (Sec 33)

3 Seat trim clip	5 Seat trim	7 Rubber diaphragm clip
4 Trim screws	6 Cushion overlap trim	8 Rubber diaphragm and switch

34 Fault diagnosis – electrical system

Symptom	Reason(s)
Starter fails to turn engine	Battery discharged Battery defective internally Battery terminal leads loose or earth lead not securely attached to body Loose or broken connections in starter motor circuit Starter motor switch or solenoid faulty Starter brushes badly worn, sticking, or brush wires loose Commutator dirty, worn or burnt Starter motor armature faulty Field coils earthed
Starter turns engine very slowly	Battery in discharged condition Starter brushes badly worn, sticking or brush wires loose Loose wires in starter motor circuit
Starter spins but does not turn engine	Pinion or flywheel gear teeth broken or worn Battery discharged
Starter motor noisy or excessively rough engagement	Pinion or flywheel gear teeth broken or worn Starter motor retaining bolts loose
Battery will not hold charge for more than a few days	Battery defective internally Electrolyte level too low or electrolyte too weak due to leakage Plate separators no longer fully effective Battery plates severely sulphated Alternator drivebelt slipping Battery terminal connections loose or corroded Alternator not charging Short in lighting circuit causing continual battery drain Alternator regulator unit not working correctly
Ignition light fails to go out, battery runs flat in a few days	Drivebelt loose and slipping or broken Alternator brushes worn, sticking, broken or dirty Alternator brush springs weak or broken Internal fault in alternator

Failure of individual electrical equipment to function correctly is dealt with alphabetically, item-by-item, under the headings listed below

Horn

Horn operates all the time	Horn push either earthed or stuck down Horn cable to horn push earthed
Horn fails to operate	Blown fuse Cable or cable connection loose, broken or disconnected Horn has an internal fault
Horn emits intermittent or unsatisfactory noise	Cable connections loose Horn incorrectly adjusted

Lights

Lights do not come on	If engine not running, battery discharged Wire connections loose, disconnected or broken Light switch shorting or otherwise faulty
Lights come on but fade out	If engine not running battery discharged Light bulb filament burnt out or bulbs or sealed beam units broken Wire connections loose, disconnected or broken Light switch shorting or otherwise faulty
Lights give very poor illumination	Lamp glasses dirty Lamps badly out of adjustment
Lights work erratically – flashing on and off, especially over bumps	Battery terminals or earth connection loose Lights not earthing properly Contacts in light switch faulty

Wipers

Wiper motor fails to work	Blown fuse Wire connections loose, disconnected or broken Brushes badly worn Armature worn or faulty Field coils faulty

Symptom	Reason(s)
Wiper motor works very slowly and takes excessive current	Commutator dirty, greasy or burnt Armature bearings dirty or unaligned Armature badly worn or faulty
Wiper motor works slowly and takes little current	Brushes badly worn Commutator dirty, greasy or burnt Armature badly worn or faulty
Wiper motor works but wiper blades remain static	Wiper motor gearbox parts badly worn

Fig. 10.21 Wiring diagram, L, HL, and HLS models

1   Panel illumination lamps
2   Direction indicator warning lamp – LH
3   Direction indicator warning lamp – RH
4   Oil pressure warning lamp
5   Ignition warning lamp
6   Choke warning lamp
7   Choke warning lamp switch
8   Brake pad warning lamp
9   Handbrake warning lamp
10   Brake failure warning lamp
11   Seat belt warning lamp
12   Driver seat belt switch
13   Passenger seat belt switch
14   Passenger seat switch – seat belt
15   Brake fluid level switch
16   Blocking diode
17   Brake pad wear sensors
18   Handbrake warning lamp switch
19   Fuel level indicator tank unit
20   Fuel level indicator
21   Coolant temperature transducer
22   Coolant temperature indicator
23   Voltage stabiliser
24   Instrument main beam warning lamp
25   Side light warning lamp
26   Horn
27   Rheostat – panel lights +
28   Ballast cable
29   Distributor
30   Ignition coil
31   Alternator
32   Starter motor solenoid
33   Battery
34   Radiator cooling fan
35   Radiator cooling fan stat
36   Oil pressure switch
37   Tailgate screen switch/warning lamp
38   Tailgate screen relay
39   Starter solenoid relay
40   Fuel pump relay
41   Fuel pump
42   Ignition switch
43   Lighting switch
44   Headlamp dip beam
45   Headlamp main beam
46   Fusebox
47   Headlamp main/dip beam/flasher, horn switch
48   Fog rear-guard switch/warning lamp

49   Side lamp – RH
50   Side lamp – LH
51   Switch illumination lamp
52   Heater control illumination
53   Cigar lighter illumination lamp
54   Glove box illumination lamp
55   Glove box illumination switch
56   Fog rear-guard lamp
57   Tail lamp – RH
58   Tail plate illumination lamps
59   Tail lamp – LH
60   Stop lamp switch
61   Stop lamps
62   Reverse lamp switch
63   Reverse lamps
64   Tailgate screen
65   Interior lamp
66   Rear interior lamp
67   Interior lamp switch
68   Luggage lamp
69   Luggage lamp switch
70   Clock
71   Windscreen wiper motor
72   Windscreen wiper/washer switch
72A Programme washer unit+
73   Windscreen washer motor
74   Direction indicator/hazard flasher unit
75   Hazard warning switch
76   Cigar lighter
77   Tailgate washer switch
78   Tailgate washer motor
79   Tailgate wiper switch
80   Tailgate wiper motor
81   Heater motor switch
82   Heater motor
83   Direction indicator switch
84   RH front direction indicator lamps
85   LH front direction indicator lamps
86   Radio/cassette player connector*
87   Window lift safety relay*
88   Window lift circuit*
89   Front foglamp switch
90   Front foglamp relay
91   Front foglamps
92   Central door locking system*
+ See Vanden Plas circuit diagram.
* See auxiliary circuit diagram.

Some components listed in this key are optional equipment on certain models.

Wire Colour Code

B	= Black	LG	= Light Green	P	= Purple	U	= Blue
G	= Green	N	= Brown	R	= Red	W	= White
K	= Pink	O	= Orange	S	= Slate	Y	= Yellow

When two colour code letters are shown on a wire the first denotes the main colour and the second denotes the tracer colour.

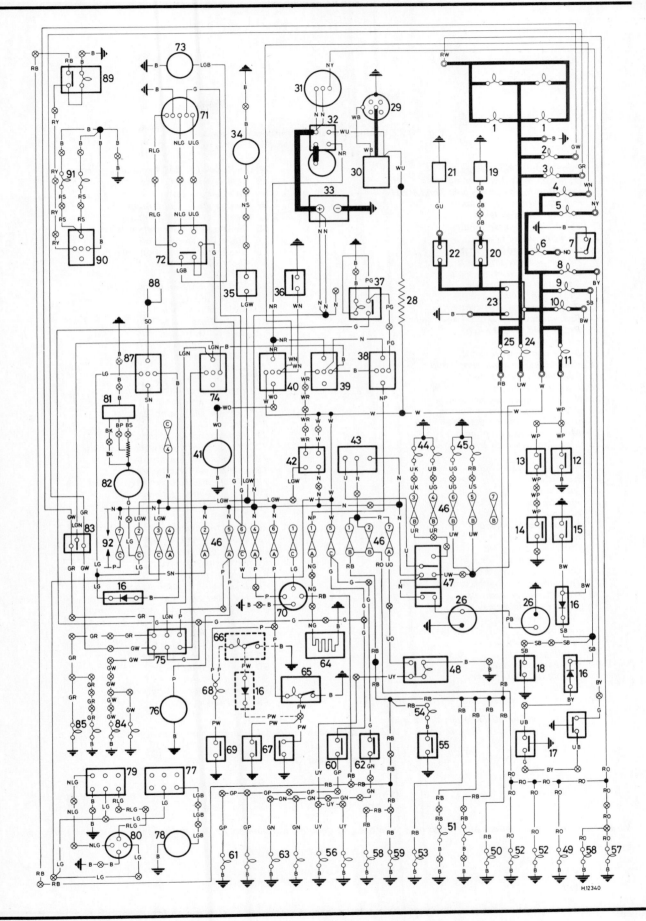

H12340

Fig. 10.22 Wiring diagram – Vanden Plas model

1   Panel illumination lamps
2   Direction indicator warning lamp – LH
3   Direction indicator warning lamp – RH
4   Oil pressure warning lamp
5   Ignition warning lamp
6   Bulb failure warning light
7   Fuel low warning light
8   Brake pad warning lamp
9   Handbrake warning lamp
10 Brake failure warning lamp
11 Seat belt warning lamp
12 Driver seat belt switch
13 Passenger seat belt switch
14 Passenger seat switch – seat belt
15 Brake fluid level switch
16 Blocking diode
17 Brake pad wear sensors
18 Handbrake warning lamp switch
19 Fuel level indicator tank unit
20 Fuel level indicator
21 Water temperature transducer
22 Water temperature indicator
23 Instrument voltage stabiliser
24 Main beam warning lamp
25 Side light warning lamp
26 Horns
27 Rheostat – panel lights
28 Ballast cable
29 Distributor
30 Ignition coil
31 Alternator
32 Starter motor solenoid
33 Battery
34 Radiator cooling fan
35 Radiator cooling fan stat
36 Oil pressure switch
37 Tailgate screen switch/warning lamp
38 Heated rear screen relay
39 Starter solenoid relay
40 Fuel pump relay
41 Fuel pump
42 Ignition switch
43 Lighting switch
44 Headlamp dip beam
45 Headlamp main beam
46 Fusebox
47 Headlamp main/dip beam/flasher, horn switch

48 Fog rear-guard switch/warning lamp
49 Side lamp – RH
50 Side lamp – LH
51 Switch illumination lamp
52 Heater control illumination
53 Cigar lighter illumination lamp
54 Glove box illumination lamp
55 Glove box illumination switch
56 Fog rear-guard lamp
57 Tail lamp – RH
58 Number plate illumination lamps
59 Tail lamp – LH
60 Stop lamp switch
61 Stop lamps
62 Reverse lamp switch
63 Reverse lamps
64 Tailgate screen
65 Interior lamp
66 Rear interior lamp
67 Interior lamp door switches
68 Luggage area lamp
69 Luggage area lamp switch
70 Clock
71 Windscreen wiper motor
72 Windscreen wiper/washer switch
72A Programme washer unit
73 Windscreen washer motor
74 Direction indicator/hazard flasher unit
75 Hazard warning switch
76 Cigar lighter
77 Tailgate washer switch
78 Tailgate washer motor
79 Tailgate wiper switch
80 Tailgate wiper motor
81 Heater motor switch
82 Heater motor
83 Direction indicator switch
84 RH front direction indicator lamps
85 LH front direction indicator lamps
86 Radio/cassette player connector*
87 Window lift safety relay*
88 Window lift circuit*
89 Front foglamp switch
90 Front foglamp relay
91 Front foglamps
92 Central door locking system*
93 Bulb failure units

*See auxiliary circuit diagram
Several of the components listed in this key are optional equipment on certain models.

Wire Colour Code

B = Black	LG = Light Green	P = Purple	U = Blue
G = Green	N = Brown	R = Red	W = White
K = Pink	O = Orange	S = Slate	Y = Yellow

When two colour code letters are shown on a wire the first denotes the main colour and the second denotes the tracer colour.

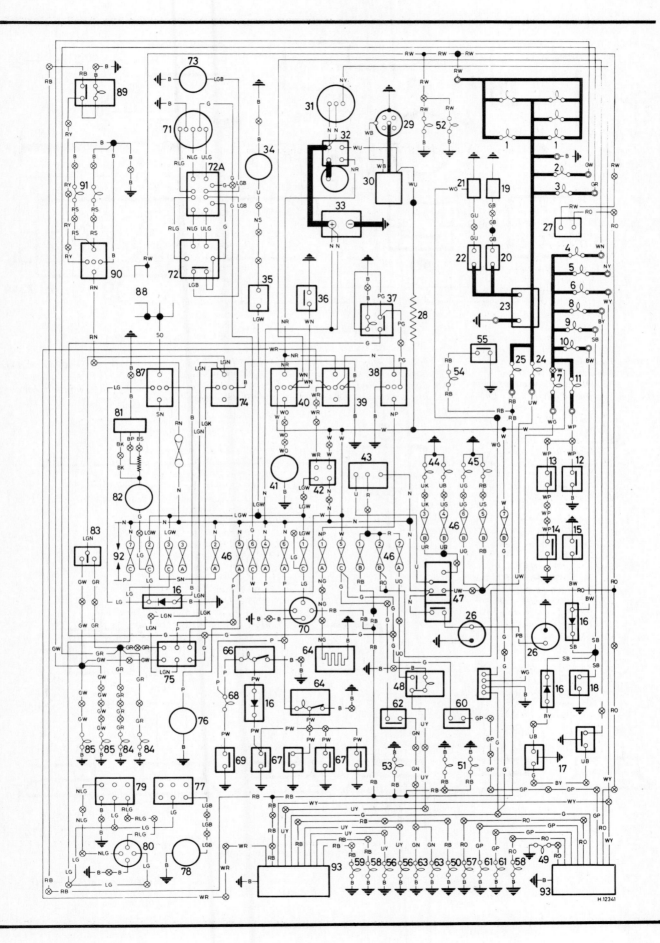

H.12341

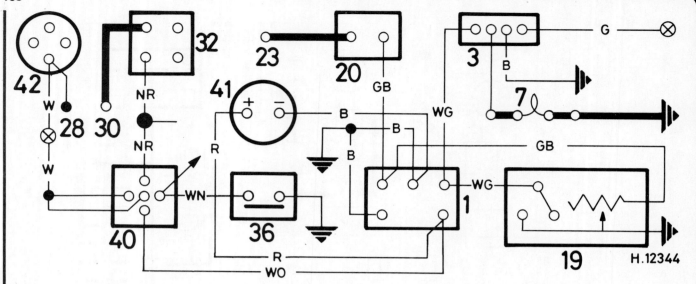

Fig. 10.23 Wiring diagram – fuel pump

Refer to main diagram

1	Connector – fuel tank	(7)	Fuel low warning light	(28)	Ballast cable	(40)	Fuel pump relay
2	Fuel low switch	(19)	Fuel level tank unit	(29)	Ignition coil	(41)	Fuel pump
3	Delay unit – fuel low warning light	(20)	Fuel level indicator	(32)	Starter motor solenoid	(42)	Ignition switch
(4)	Oil pressure warning light	(23)	Instrument voltage stabiliser	(36)	Oil pressure switch		

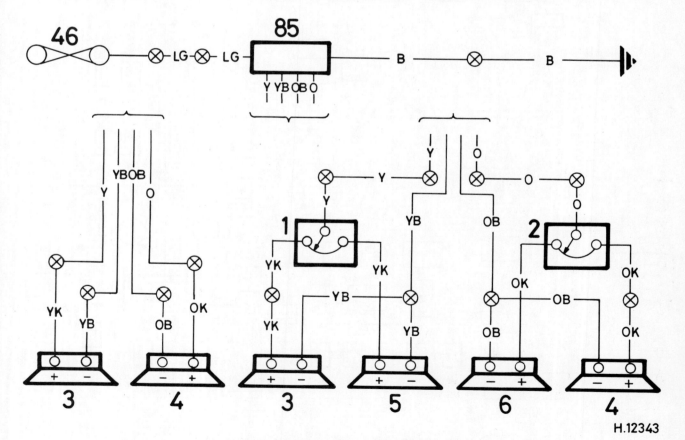

Fig. 10.24 Wiring diagram – radio/cassette player

1	Balance control – LH	3	Speaker – front – LH	5	Speaker – front – RH
2	Balance control – RH	4	Speaker – rear – LH	6	Speaker – rear – RH

Refer to main diagram
(46) Fusebox
(85) Radio/cassette player

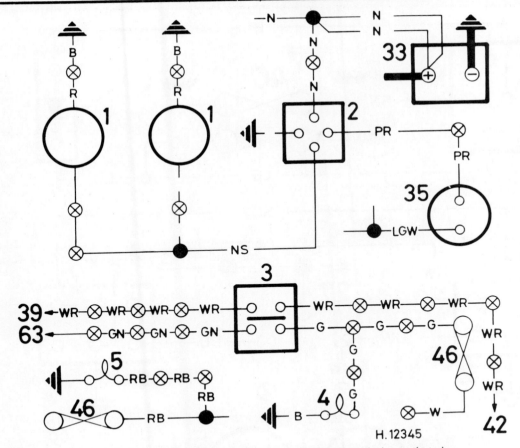

Fig. 10.25 Wiring diagram – automatic transmission wth twin cooling fans

1	Cooling fan motors	4	Selector indicator lamp	**Refer to main diagram**		(42)	Ignition coil
2	Relay – cooling fan motors	5	Selector illumination lamp	(33)	Battery	(46)	Fusebox
3	Starter inhibitor/reverse lamp switch			(35)	Radiator cooling fan stat	(63)	Reverse lamps
				(39)	Starter solenoid relay		

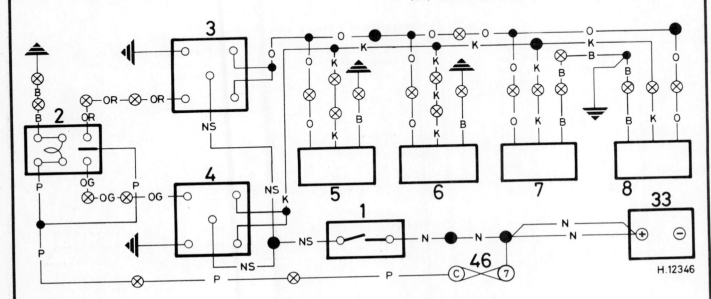

Fig. 10.26 Wiring diagram – central door locking

1	Thermal circuit breaker	3	Open relay	**Door lock solenoids**		7	Rear door – RH
2	Door lock solenoid switch	4	Close relay	5	Front door – LH	8	Tail door
				6	Rear door – LH		

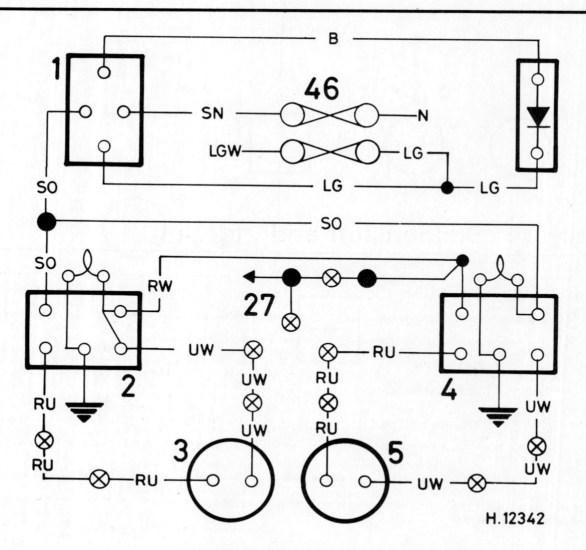

Fig. 10.27 Wiring diagram – electric window lift

1   Relay front window lift
2   Window lift switch – LH
3   Window lift motor – LH

4   Window lift switch – RH
5   Window lift motor – RH

Refer to main diagram
(27) Rheostat – Panel lights
(46) Fusebox

Chapter 11 Suspension and steering

Contents

Specifications

Front suspension

Type	Hydragas independent interconnected to rear
Camber	0°30′ ± 0°30′ positive
Castor angle	1°30′ ± 0°30′ positive
Steering axis inclination	11°
Wheel alignment	Parallel ± 1.6 mm

Rear suspension

Type	Hydragas independent interconnected to front
Camber	1° ± 0°30′ negative
Toe-in	0.32 ± 0.28 in (8.0 ± 7.0 mm)

Steering

Type	Power-assisted rack and pinion
Turning circle (between kerbs)	37 ft 10 in (11.53)
Number of turns of steering wheel (lock to lock)	3.26
Rack lubricant type/specification	Multi-purpose lithium based grease to NLGI No 2 (Duckhams LB 10)
Reservoir fluid type/specification:	
Up to VIN 122690	Multigrade engine oil, viscosity SAE 15W/50, to API SE/SF (Duckhams Hypergrade)
From VIN 122691	Automatic transmission fluid to M2C 33G or M2C 33F (Duckhams Q-Matic)
Fluid capacity	2.01 Imp pts (1.1 litre)

Torque wrench settings

	lbf ft	Nm
Front suspension		
Hub nut	200	272
Bump rubber to bracket	30	41
Bumper rubber bracket to valance	15	21
Suspension arm balljoint pin nuts	40	55
Suspension arm balljoint housings	70	97
Upper arm pivot bolt	120	163
Lower arm pivot bolt	50	69
Lower arm rear bush to body	15	21

Rear suspension

Cross-tube mounting bolts:		
To body	37	51
To cross-tube	20	28
Radius arm pivot bolt	120	163
Rebound strap bracket bolts	44	61
Bump rubber to radius arm nut	20	28
Reaction rubber to body screws	20	28

Steering

Steering wheel nut	35	48
Steering arm to hub bolts:		
Plain	35	48
Dimpled	45	62
Tie-rod end balljoint taper pin nut	35	48
Steering rack to trunnion bolts	38	53
Rack trunnion to body bolts	20	28
Intermediate shaft pinch-bolt	20	28
Power steering reservoir setscrew	35	48
Tie-rod end locknuts	35	48

1 General description

The suspension system is independent on all four wheels utilising the Hydragas system. The fluid chambers in the front and rear Hydragas units on the same side of the car are interconnected. Each Hydragas displacer unit consists of a nitrogen-filled spherical chamber and a displacer chamber between which is located a two way valve to provide the necessary suspension damping. The space above the flexible separator in the upper chamber is charged with nitrogen to provide the springing effect. The space between the separator and the diaphragm is fitted with fluid and is at a higher pressure than that of the nitrogen. This pressure differential has the effect of compressing the nitrogen and lifting the separator from the bottom of the spherical chamber and causing the fluid and gas pressures to equalise. When a front wheel is on a ridge and a rear wheel is in a hollow, the compression of the front diaphragm in the Hydragas unit displaces fluid which is transferred through the interconnecting pipe to the rear displacer chamber on the same side of the car. As this displaced fluid is accommodated by the opening of the rear displacer chamber, the fluid pressure in the system will not overcome the damper valve and so the nitrogen springs are not deflected. Movement of the body is therefore restricted to a minimum.

When a front wheel hits a bump with the rear wheel still on level ground, fluid is displaced and both front and rear nitrogen springs are displaced.

When both front and rear roadwheels hit bumps simultaneously, the compression of the front and rear diaphragms increases the fluid pressure making it pass through the damper valves, thus compressing the nitrogen springs. Under these conditions, fluid flow through the interconnecting pipe is minimal. The fluid used in the Hydragas suspension system is of a special mixture and no substitute must be used.

The Hydragas displacer units transmit their action in the case of the front wheels, through upper and lower suspension arms which are connected to the swivel hub of the driveshaft. With the rear units, movement is transmitted through radius arms attached to cross-tubes.

The steering is of rack-and-pinion, power-assisted type. The steering column incorporates an intermediate shaft fitted with two universal joints.

For details of hub bearings, refer to Chapter 8.

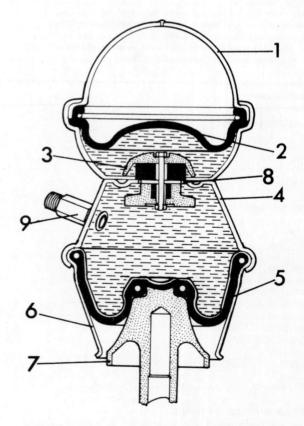

Fig. 11.2 Sectional view of Hydragas displacer unit (Sec 1)

1	Nitrogen filled chamber	6	Displacer strut
2	Separator	7	Piston
3	Two-way damper valve	8	Bleed hole
4	Displacer chamber	9	Adaptor
5	Diaphragm		

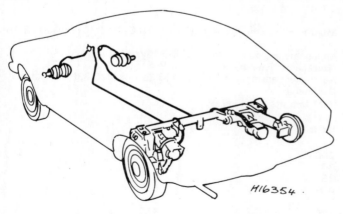

H16354

Fig. 11.1 Hydragas suspension system (Sec 1)

2.2 Steering rack greaser

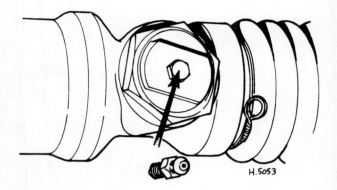

Fig. 11.3 Steering rack grease plug (Sec 2)

2 Suspension and steering – maintenance

1   No maintenance is required to the Hydragas suspension system.
2   At the intervals specified in Routine Maintenance, remove the plug from the top of the steering rack pinion housing and screw in a grease nipple instead. Apply a grease gun and give five or six strokes only. Remove the grease nipple and refit the plug (photo). Note that later models are fitted with a 'sealed for life' steering rack which requires no periodic greasing, and the plug is no longer fitted.
3   Keep the power steering pump drivebelt in good condition and correctly tensioned as described in Chapter 2, Section 10.
4   Regularly check the fluid level in the power steering fluid reservoir. Do this first thing in the morning when the fluid is cold and the engine switched off.
5   Twist the reservoir cap and remove it.
6   Wipe the dipstick which is attached to the cap and refit the cap. Remove the cap for the second time and read off the fluid level. Add oil or fluid as specified to bring the level between the end of the dipstick and the Full Cold mark.
7   Power steering, fitted up to VIN 122690, should be filled with engine oil, viscosity SAE 15W/50, to API SE/SF (Duckhams Hypergrade). The steering fitted from VIN 122691 should be filled with automatic transmission fluid to M2C 33G or M2C 33F (Duckhams Q-Matic).
8   At the intervals specified in Routine Maintenance check all suspension and steering joints and bushes for wear.
9   Enlist the help of an assistant to check for wear. While he is turning the steering wheel a few degrees in each direction inspect the tie-rod end balljoints. Any movement or free movement in the steering will indicate wear in the balljoints or possibly wear in the steering rack.
10 Check the front wheel alignment at the recommended intervals (see Section 26).

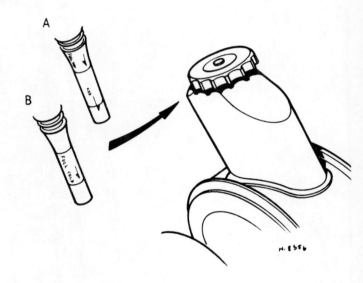

Fig. 11.4 Power steering pump reservoir dipstick markings (Sec 2)

A   Fluid hot   B   Fluid cold

3 Suspension overhaul – precautions

1   Prior to dismantling any part of the suspension system, if such work will entail disconnection of the Hydragas system, then it will have to be first depressurised, then recharged, after the work has been completed. This can only be carried out by your Austin-Rover dealer or a service station having the necessary equipment. On no account touch the Hydragas system valves or the system will be rendered inoperative or at least, the trim level upset (See Section 16).
2   In order that suspension overhaul work can be carried out at home, it is quite safe to drive from and to your nearest dealer with the system depressurised, provided the road surface is reasonable and the road speed is kept below 30 mph (50 km/h). The action of the suspension bump rubbers will provide the necessary temporary suspension characteristics.

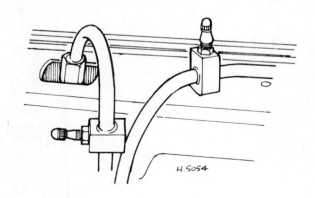

Fig. 11.5 Hydragas pipeline valves on engine compartment rear bulkhead (Sec 3)

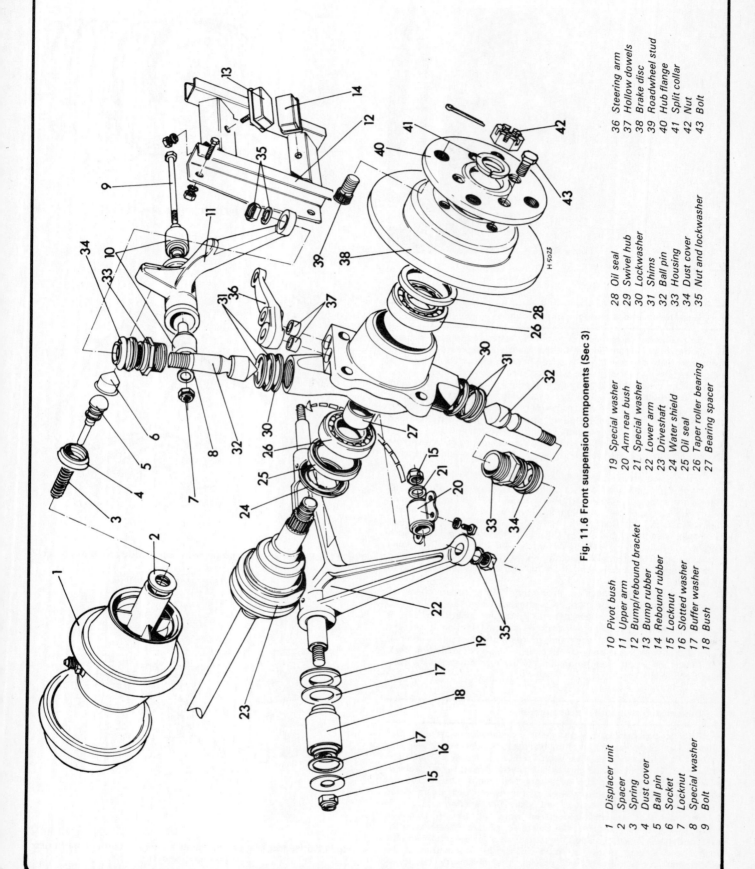

Fig. 11.6 Front suspension components (Sec 3)

1 Displacer unit
2 Spacer
3 Spring
4 Dust cover
5 Ball pin
6 Socket
7 Locknut
8 Special washer
9 Bolt

10 Pivot bush
11 Upper arm
12 Bump/rebound bracket
13 Bump rubber
14 Rebound rubber
15 Locknut
16 Slotted washer
17 Buffer washer
18 Bush

19 Special washer
20 Arm rear bush
21 Special washer
22 Lower arm
23 Driveshaft
24 Water shield
25 Oil seal
26 Taper roller bearing
27 Bearing spacer

28 Oil seal
29 Swivel hub
30 Lockwasher
31 Shims
32 Ball pin
33 Housing
34 Dust cover
35 Nut and lockwasher

36 Steering arm
37 Hollow dowels
38 Brake disc
39 Roadwheel stud
40 Hub flange
41 Split collar
42 Nut
43 Bolt

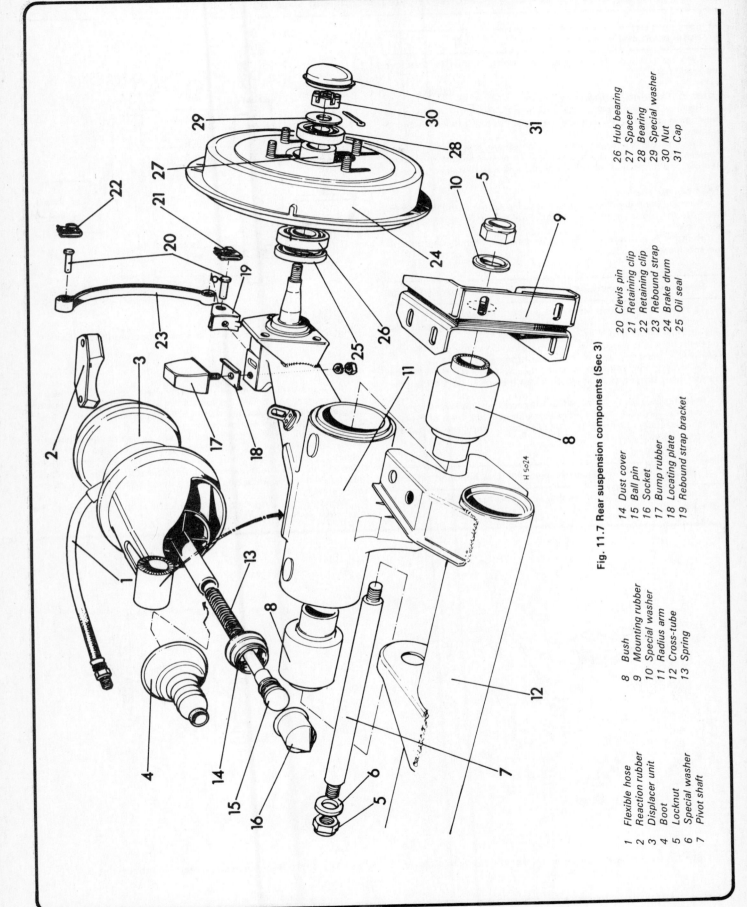

Fig. 11.7 Rear suspension components (Sec 3)

1 Flexible hose
2 Reaction rubber
3 Displacer unit
4 Boot
5 Locknut
6 Special washer
7 Pivot shaft

8 Bush
9 Mounting rubber
10 Special washer
11 Radius arm
12 Cross-tube
13 Spring

14 Dust cover
15 Ball pin
16 Socket
17 Bump rubber
18 Locating plate
19 Rebound strap bracket

20 Clevis pin
21 Retaining clip
22 Retaining clip
23 Rebound strap
24 Brake drum
25 Oil seal

26 Hub bearing
27 Spacer
28 Bearing
29 Special washer
30 Nut
31 Cap

4 Suspension bump rubbers – renewal

1   Jack up and support the appropriate side of the car and remove the roadwheel.

Front bump and rebound rubbers
2   Remove the setscrew which is secured by the nut on the inside of the engine compartment valance and which in turn retains the bump bracket.
3   Remove the three setscrews which secure the bracket to the valance and then slide the bracket along the suspension upper arm. Remove the bump and rebound rubbers which are retained by pegs.

Rear bump rubber
4   Remove the nut which secures the bump rubber to the radius arm. Separate the rubber from the locating plate noting the position of the plate lips.
5   Refitting is a reversal of removal but tighten all nuts and bolts to the specified torque.

5 Rear suspension rebound strap and reaction rubber – renewal

1   To renew the rebound strap, jack-up both the car and the radius arm so that any tension is released from the strap. Remove the spring clips and the clevis pin.
2   Refitting is a reversal of removal but the side of the strap marked REAR FACE must be towards the rear of the car.
3   To renew the reaction rubber, the system will have to be depressurised on one side. The rubber can then be unbolted from the floor.
4   Refitting the new reaction rubber is a reversal of removal but make sure that the flat side of the rubber faces towards the front of the car. Tighten the retaining bolts to the specified torque and have the system recharged by your dealer.

6 Front suspension upper arm – removal, overhaul and refitting

1   Have your dealer depressurise the appropriate side of the Hydragas system.

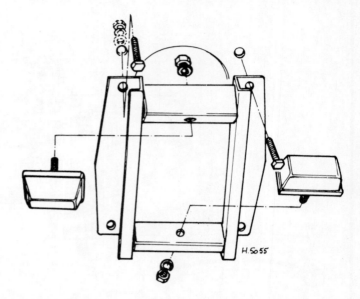

Fig. 11.8 Front suspension bump and rebound rubbers (Sec 4)

2   Raise the car and support it securely. Remove the roadwheel.
3   Disconnect the swivel balljoint from the suspension upper arm using an extractor or wedges. As the balljoint is disconnected, support the suspension under the lower arm.
4   Remove the bump and rebound bracket, as described in Section 4.
5   Remove the locknut and special washer from the end of the suspension arm pivot bolt which projects into the engine compartment. Tap the threaded end of the pivot bolt with a soft-faced mallet so that its opposite end can be gripped and withdrawn into the car interior.
6   Withdraw the suspension arm complete with knuckle joint. The latter may be levered out if required.

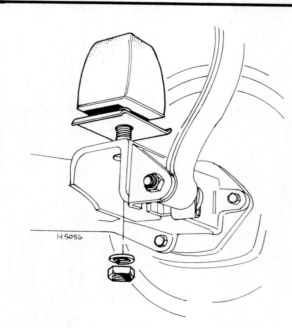

Fig. 11.9 Rear suspension bump rubber and rebound strap (Sec 5)

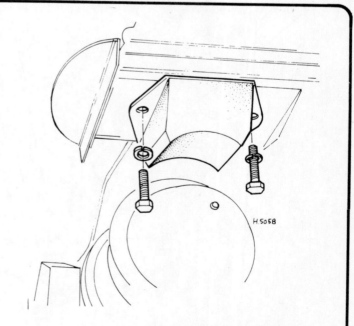

Fig. 11.10 Rear suspension reaction rubber (Sec 5)

7   Renewal of the suspension arm bushes is best left to your dealer but where pressing facilities are available, press them out using a suitable mandrel. During this operation, the flexible bushes will probably separate from the outer sleeves as pressure is applied to the inner sleeves. The outer sleeves can be removed afterwards after splitting them with a sharp cold chisel. Press the new bushes into the suspension arm by applying pressure to their outer sleeves. When the end faces of the inner sleeves are together in the centre of the arm, then there should be a projection of the outer sleeve on each side of the arm as shown in Fig. 11.12.

8   Commence refitting the suspension upper arm by packing the knuckle joint with specified grease.

9   Apply graphite grease to the knuckle joint spigot.

10 Refit the suspension arm but only tighten the pivot bolt locknut finger-tight.

11 Jack-up the suspension until the upper arm is horizontal and then tighten the pivot bolt locknut to the specified torque.

12 On completion, have your dealer recharge the Hydragas system.

---

7   Front suspension lower arm – removal, overhaul and refitting

1   Raise the front of the vehicle and remove the roadwheel.

2   Slacken both pivot bush locknuts. Remove the nut and spring washer from the lower balljoint taper pin and then using a suitable extractor, disconnect the balljoint from the lower arm.

3   Unscrew the locknut and slotted washer from the front pivot bush.

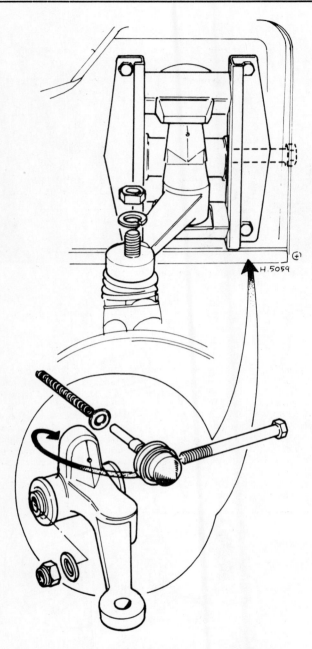

Fig. 11.11 Front suspension upper arm (Sec 6)

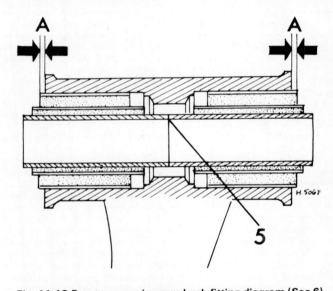

Fig. 11.12 Front suspension arm bush fitting diagram (Sec 6)

5   Arm centre line   A   0.068 to 0.078 in (1.72 to 1.97 mm)

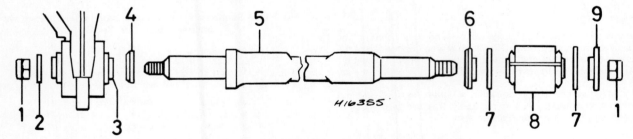

Fig. 11.13 Components of front suspension lower arm (Sec 7)

1 Locknut	4 Chamfered washer	7 Buffer washer
2 Slotted washer	5 Lower arm	8 Rear pivot bush
3 Front pivot bush	6 Inner buffer locating washer	9 Outer buffer locating washer

4   Support the lower arm and remove the screws which hold the rear pivot bush to the body.

5   Withdraw the lower arm from the front pivot bush and remove it from the car.

6   Remove the rear pivot bush locknut and the outer buffer locating washer, the buffer, rear pivot bush, inner buffer and locating washer.

7   Renewal of the suspension lower arm front bush can be carried out using a length of threaded studding, two nuts, washers and suitable distance pieces to withdraw it. Pull the new bush into position until its face is flush with the mounting bracket. The rear pivot bush is renewed as an assembly.

8   Refitting is basically a reversal of removal, but observe the following points.

9   Note that the chamfered edge of the inner buffer washer is towards the lower arm.

10   Make sure that all the components are fitted in their correct sequence. The locknuts should only be screwed on finger tight at this stage.

11   Insert the lower arm into the front pivot bush, fit the slotted washer and screw on the locknut finger tight.

12   Fit the rear pivot bush fixing screws finger tight.

13   Connect the lower arm to the bottom balljoint. Fit the spring washer and taper pin nut.

14   Locate a jack under the suspension lower arm and raise it until the arm is horizontal. Now tighten all nuts and screws to the specified torque.

8.2 Disconnecting upper swivel balljoint

8   Front suspension swivel balljoints – removal and refitting

1   Raise the front of the car, support it adequately and remove the front roadwheel. Unbolt the brake caliper and tie it up out of the way.

2   *If the upper swivel balljoint* is to be removed, place a supporting jack under the suspension lower arm and then remove the ball pin nut and washer and disconnect the balljoint from the suspension arm using a separator or two wedges (photo).

3   Lower the jack under the suspension arm and disconnect the balljoint from the upper arm.

4   Flatten the locktab and unscrew the balljoint housing from the swivel hub.

5   *If the lower swivel balljoint* is to be removed, then first slacken the suspension lower arm pivot locknuts, otherwise the operations are similar to those described for the upper joint.

6   Before fitting a new balljoint to the swivel hub, thoroughly clean out the seating in the hub. Fit the new balljoint to the hub but not including the locking washer or any shims. Tighten until the ball pin is just nipped but is still free to swivel without any endplay. Now measure the gap (A) between the housing and the swivel hub using feeler blades. From this dimension subtract between 0.009 and 0.013 in (0.23 to 0.33 mm). The resulting dimension must be equalled by the thickness of the locking washer and shim pack to provide the balljoint with the necessary preload. To meet this requirement, shims are available in the following thicknesses:

0.002 in (0.05 mm)
0.003 in (0.08 mm)
0.005 in (0.13 mm)
0.010 in (0.25 mm)
0.030 in (0.75 mm)

7   Assemble the shims and locking washer and then screw in the balljoint swivel housing to the specified torque. Bend up the tab of the locking washer. Reconnect and tighten the ball pin nuts.

8   Tighten the suspension lower arm nuts when the weight of the car is on the arm.

9   If difficulty is experienced in loosening or tightening a ball-stud nut due to the ball-stud (taper pin) turning, apply pressure with a jack or long lever to force the taper pin into its seat.

9   Front displacer unit – removal and refitting

1   Have your dealer depressurise the Hydragas system.

2   Remove the front suspension upper arm (Section 6).

3   Unscrew the interconnecting pipe union from the displacer unit.

4   Withdraw the displacer unit from its location in the body.

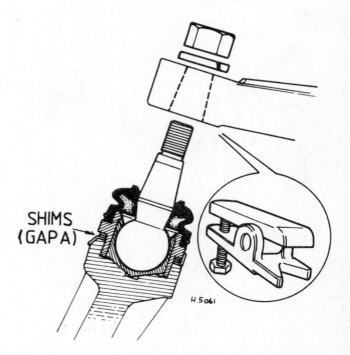

SHIMS
(GAP A)

H.5061

Fig. 11.14 Sectional view of suspension upper balljoint with balljoint separator (Sec 8)

5   The knuckle joint can be levered from the upper arm.

6   Refitting is a reversal of removal but observe the following points:

    (a)   Pack the knuckle joint with specified grease

    (b)   Ensure that the interconnecting pipe is centrally placed in the body aperture after it is connected

7   Have your dealer re-charge the Hydragas system.

10   Rear displacer unit – removal and refitting

1   Have your dealer depressurise the Hydragas system on the side of the car being overhauled (photo).

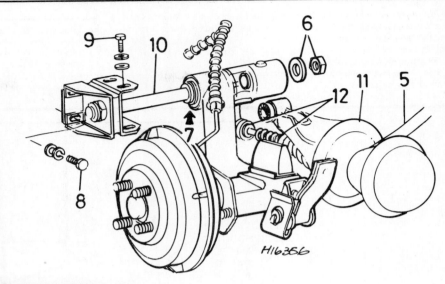

Fig. 11.15 Rear displacer unit (Sec 10)

5   Hydraulic hose
6   Locknut and washer
7   Radius arm
8   Crosstube mounting
    screws
9   Crosstube mounting
    screws
10 Pivot shaft
11 Displacer unit
12 Knuckle joint and spring

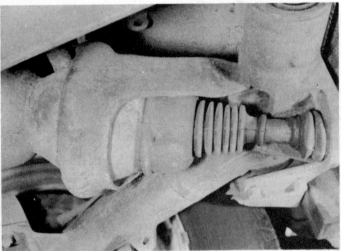

10.1 Rear Hydragas system

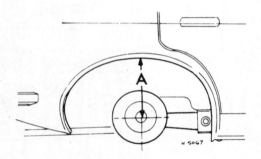

Fig. 11.16 Rear suspension radius arm setting diagram (Sec 10)

A   11.2 in (285.0 mm)

2   Jack-up the rear of the car, support it securely and remove the roadwheel.
3   Disconnect the flexible hose at its union with the rigid interconnecting pipe.
4   Remove the locknut and special washer from the inner end of the radius arm pivot shaft and support the radius arm assembly.
5   Remove the two setscrews which secure the cross-tube mounting rubber to the body, and the four screws securing the mounting rubber to the cross-tube. The jack may have to be lowered slightly in order to obtain access to the front setscrew. Withdraw the mounting rubber and pivot shaft from the radius arm so that the displacer can be released.
6   Refitting is a reversal of removal but observe the following points:

Initially only tighten the mounting rubber to body and cross-tube screws finger-tight, then support the radius arm in such a position that dimension 'A' is as specified in Fig. 11.16. Now tighten the mounting rubber screws to the specified torque.

7   Have your dealer re-charge the Hydragas system.

11 Rear displacer knuckle joint and boot – removal and refitting

1   Have your dealer depressurise the Hydragas system.
2   Jack up the rear of the car and support it securely on axle stands.
3   Hold the radius arm in the full rebound position.

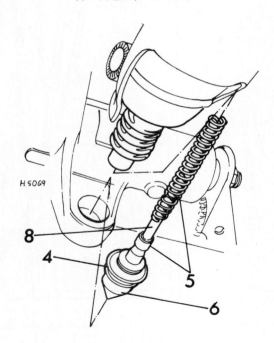

Fig. 11.17 Rear displacer unit knuckle joint assembly (Sec 11)

4   Boot   6   Cup
5   Knuckle ball and spring 8   Balljoint spigot

4 Pull the boot from the lip of the cup, pull the ball from the cup and extract the knuckle ball spring from the displacer strut. Extract the cup from the bore of the radius arm.
5 Pull the boot from the displacer unit and strut.
6 Refitting is a reversal of removal.
7 Have the Hydragas system re-charged and the body trim height adjusted.

12 Displacer rigid connecting pipe – removal and refitting

1 Have your dealer depressurise the Hydragas system.
2 Working beneath the car, disconnect the pipe union at the front displacer unit.
3 *If the left-hand displacer pipe is being removed, remove the*

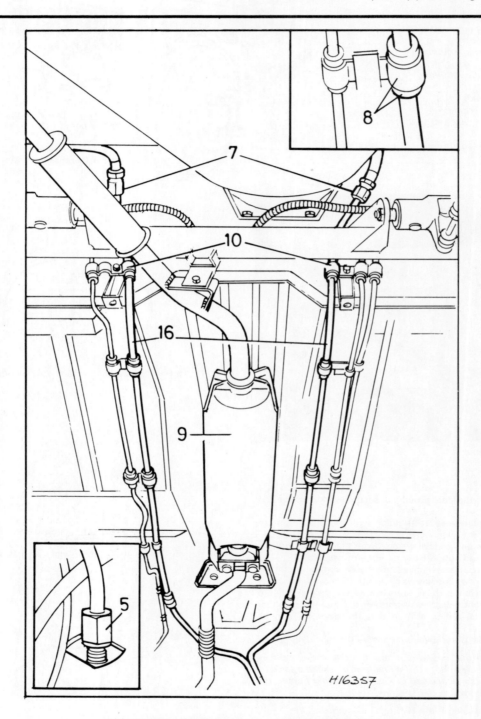

Fig. 11.18 Displacer connecting pipelines (Sec 12)

5 Pipe union	8 Pipe clips and insulators	10 Pipe clip and
7 Pipe union	9 Exhaust silencer	insulator
		16 Pipelines

exhaust pipe clamp. Remove the rear roadwheel and support the radius arm in its normal static position, using a jack positioned under the brake drum. The exhaust pipe can now be removed.

4 *With either pipe,* disconnect the union from the rear displacer flexible hose. If a restrictor washer is fitted between the pipe and the hose, discard it if a new pipe is to be installed.

5 Support the cross-tube mounting with a jack and remove the two screws which secure the mounting bracket to the body. Now lower the crosstube.

6 Unclip the displacer from the underside of the body and remove the rubbers.

7 Detach any pipe clips and pull the displacer connecting pipe down from between the cross-tube and the body. With the help of an assistant, withdraw the pipe from the engine compartment, passing it between the steering rack and the body.

8 Refitting the pipe is a reversal of removal but only use the restrictor washer if the original pipes are being fitted.

9 Have the Hydragas system recharged and the body trim height adjusted by your dealer.

13 Displacer flexible hose – general

The flexible hose fitted to the rear displacer unit is an integral part, and cannot be removed from the displacer. If removing the displacer, the hose should only be disconnected at its union with the rigid interconnecting pipe.

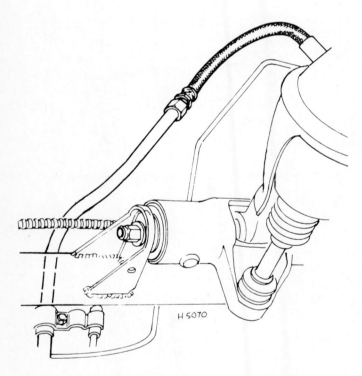

H 5070

Fig. 11.19 Rear displacer unit flexible hose (Sec 13)

14 Rear suspension radius arm – removal, overhaul and refitting

1 Have your dealer depressurise the Hydragas system.
2 Jack up the rear of the car and support it on stands.
3 Disconnect the flexible hose at the rigid pipe union. Plug the open end of the hose.
4 Disconnect the brake pipe and flexible hose at the radius arm bracket. Plug the hose.

5 Remove the brake drum, as described in Chapter 9.
6 Remove the strap which retains the handbrake cable to the radius arm.
7 Remove the three nuts and their special bolts which secure the brake backplate to the radius arm.
8 Withdraw the complete brake assembly, pass it over the radius arm and support it on blocks underneath the body.
9 Remove the locknut and special washer from the inner end of the pivot shaft.
10 Support the centre of the radius arm, remove the spring clip and clevis pin and detach the rebound strap from the radius arm bracket.
11 Remove the setscrews which secure the mounting rubber to the body, also the four setscrews which secure it to the cross-tube.
12 Release the mounting rubber and the pivot shaft from the mounting tube brackets and withdraw the radius arm. Detach the mounting rubber and pivot shaft from the radius arm.
13 The displacer unit, the knuckle joint spring, the knuckle joint, the rebound strap bracket and the bump rubber can all be removed from the radius arm if required.
14 The radius arm bushes can be renewed as described for front suspension arms in Section 6, paragraph 7. Make sure that the new bushes are pressed in flush with the radius arm.
15 Refitting is a reversal of removal but apply graphite grease to the knuckle joint spigot and screw the inner pivot nut and mounting rubber screws finger tight until the radius arm has been jacked-up to provide a dimension 'A' as shown in Fig. 11.16. All nuts, screws and bolts should then be tightened to their specified torque wrench settings.
16 Bleed the brake hydraulic circuit and have the Hydragas system recharged and the body trim height checked and adjusted by your dealer.

15 Rear suspension cross-tube and mountings – removal and refitting

1 Have your dealer depressurise the Hydragas system.
2 Jack-up the rear of the car, support it securely and remove the rear roadwheels.
3 Disconnect the exhaust pipe front strap from the cross-tube mounting.
4 Disconnect the exhaust pipe rear strap from the tailpipe and support the pipe.
5 Disconnect the displacer flexible hoses from the rigid pipe unions.
6 Remove the locknuts and special washers from the inner ends of the pivot shafts.
7 Release the handbrake cables from their clips on the cross-tube.
8 On one side of the car remove the screws which secure the cross-tube mountings to the body and to the cross-tube itself.
9 Free the mounting rubber and pivot shaft from the cross-tube bracket.
10 Move the radius arm on one side of the car clear of the cross-tube and support the arm on a block or axle stand.
11 Support the cross-tube at its centre point, preferably on a trolley jack.
12 Repeat the operations described in paragraphs 8, 9 and 10 on the opposite side of the car.
13 Lower the cross-tube jack and withdraw the assembly from beneath the car.
14 Refitting is a reversal of removal, but before tightening the cross-tube mounting nuts and the radius inner pivot nuts, set the radius arm as shown in Fig. 11.16. Hold the displacer units hard against their reaction rubber pads while tightening the radius arm inner pivot nuts.
15 Have the Hydragas recharged and the body trim height checked by your dealer.

16 Body trim height – adjustment

1 To ensure correct operation of the suspension, good roadholding and positive steering, it is important that the correct body trim height is maintained at all times.
2 This is a job for your dealer who will have the necessary equipment to increase or decrease the pressure in the Hydragas system using the two valves.
3 When the car is standing (without occupants) on a level surface

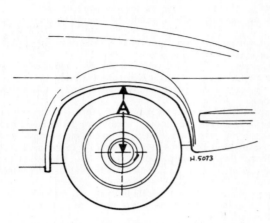

Fig. 11.20 Body trim height diagram (Sec 16)

A   14.5 ± 0.25 in (368.4 ± 6.0 mm)
Variation between the two sides must not exceed 0.39 in (10.0 mm)

17.2 Disconnecting tie-rod balljoint

with tyres correctly inflated, the body trim height is correct when the dimension 'A' measured between the hub centre and the wheel arch is as specified in Fig. 11.20.

17 Tie-rod ends – removal and refitting

1   Where 'lost motion' is observed at the steering wheel or when movement or shake can be felt at the tie-rod end balljoint if the tie-rod is gripped and moved up and down, then the tie-rod end must be renewed. Unscrew the ball stud nut.

2   Disconnect the tie-rod end ball pin from the eye of the steering arm using a suitable extractor or forked wedges (photo).

3   Holding the tie-rod end quite still in an open ended spanner release the locknut. Unscrew the tie-rod end and count the number of threads between the end of the tie-rod and the face of the locknut to give an approximate setting position on refitting.

4   Screw on the new tie-rod and reconnect it to the steering eye. Check the front wheel alignment, as described in Section 26.

5   Should the steering arms be removed, note the locating dowels and tap the arms upwards to release them after withdrawing the bolts.

6   If difficulty is experienced in loosening or tightening a ball-stud nut, due to the ball-stud (taper pin) turning, apply pressure with a jack or long lever to force the taper pin into its seat.

18 Steering gear bellows – renewal

1   Should the bellows on the rack housing split or become perished, they must be renewed. If the bellows have been in a faulty condition for some time then the rack assembly should be removed, dismantled, cleaned and lubricated with fresh grease.

2   To remove the bellows, first disconnect the tie-rod end balljint pins from the eyes of the steering arms, then holding the tie-rod end quite still in an open-ended spanner, release the locknut.

3   Unscrew the tie-rod end and count the number of threads between the end of the tie-rod and the face of the locknut. Record this as it will give an approximate position for refitting of the tie-rod ends during reassembly.

4   Cut the retaining wires or remove the clips from the bellows and remove them from the tie-rods.

5   Refitting is a reversal of removal but on completion, check the front wheel alignment (Section 26).

19 Steering wheel – removal and refitting

1   Set the steering in the straight-ahead position and prise the crash pad from the steering wheel spokes (photo).

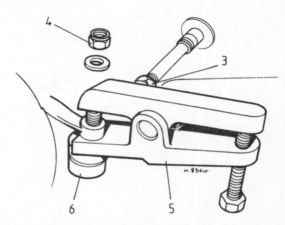

Fig. 11.21 Separating tie rod end balljoint from steering arm (Sec 17)

3 Locknut	5 Tool
4 Nut	6 Balljoint

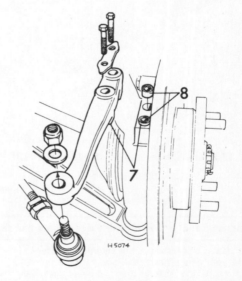

Fig. 11.22 Attachment of steering arm to swivel hub showing hollow locating dowels (7 and 8) (Sec 17)

19.1 Steering wheel crash pad

19.2 Interior of steering column shroud

19.3 Steering wheel nut

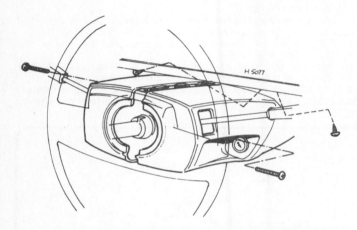

Fig. 11.23 Steering column cowl attachment (Sec 19)

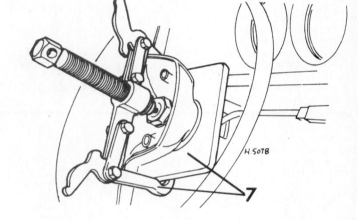
Fig. 11.24 Using a puller (7) to remove the steering wheel (Sec 19)

2   Remove the two securing screws and withdraw the left-hand cowl. Remove the right-hand cowl. If the cowls are difficult to remove, remove them after the steering wheel has been withdrawn (photo).
3   Turn the ignition key to unlock the column lock and then unscrew and remove the steering wheel nut and washer (photo).
4   Mark the relationship of the steering wheel to the steering shaft and then screw the steering wheel nut on again but only two or three threads.
5   If the steering wheel will not pull off by hand pressure, use a suitable two or three-legged puller but protect the lower edge of the steering wheel hub by inserting a thin wooden or metal plate to act as a leverage point for the claws of the puller.
6   Refitting is a reversal of removal but make sure that the slots and dog of the switch bush engage correctly with the switch and wheel hub. Align the wheel to the shaft using the marks made before removal with the roadwheel in the straight-ahead position.
7   Tighten the steering wheel nut to the specified torque.

20 Steering column lock/switch – removal and refitting

1   Remove the parcels tray under cover and disconnect the multi-connector from the ignition switch and release the cable retaining clip.
2   Release the steering column cowls and move them to each side.
3   Centre-punch the heads of the shear type bolts which secure the lock assembly to the steering column. Drill out the bolts completely or drill holes to accept bolt extractors.
4   Remove the lock and clamp plate.
5   The ignition/starter switch can be removed from the column lock after withdrawal of the small retaining screw.
6   When fitting the new lock assembly, centralise the lock body over

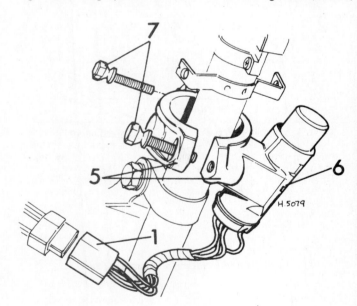
Fig. 11.25 Steering column lock (Sec 20)

1 Ignition switch connector	6 Ignition switch
5 Lock assembly	7 Shear bolts

the slot in the outer column, fit the clamp plate and screw in the shear bolts only a little more than finger-tight. Check the operation of the lock and switch at all key positions and then tighten the bolts until their heads break off.

21 Steering column – removal, overhaul and refitting

1   Withdraw the left-hand steering column cowl and tie it to one side.
2   Remove the pinch-bolt which secures the steering shaft to the intermediate shaft.
3   Disconnect the steering column lower clamp by slackening one clamp screw and removing the other one.
4   Disconnect the combination switch connector plugs.
5   Unscrew and remove the column upper clamp bolt, extract the bottom bracket from the bolt and withdraw the column complete with steering wheel into the car interior.
6   Remove the steering wheel, combination switch and lock assembly.
7   Grip the steering column carefully in a vice fitted with jaw protectors.
8   Pull the inner shaft from the top of the outer column. The bush at the base of the column will be destroyed during this operation.
9   Extract the top bush and then prise up the retaining tag and withdraw the bottom bush.
10 Renew the bushes and any other worn components.
11 Apply graphite grease to the grooves of the new steering column bushes.
12 Insert the shaft into the column so that it projects about 3 in (76.0 mm) from the lower end of the column.

13 Open the bottom bush so that it will pass over the splines of the shaft and then drive it into the outer column (chamfered end first). Make sure that one of the projections of the bush engages in the slot in the outer column. Bend down the bush retaining tag but do not let it foul the inner shaft.
14 Fit the column upper bush in a similar manner to the lower one.
15 Finally adjust the position of the shaft so that it projects 3.45 in (88.0 mm) above the outer column.

22 Steering gear – removal and refitting

1   Remove the pinch-bolt which secured the intermediate shaft to the pinion on the rack housing.
2   Raise the front of the car and support it securely.
3   Remove the front roadwheels.
4   Remove the suspension lower arm (Section 7) from the steering column side.
5   Disconnect the tie-rod ends from the steering arms.
6   Disconnect the fluid return hose from the rack pipe.
7   Disconnect the feed hose union, catching the small amount of oil which will be released.
8   Remove the bolt which secures the rack mounting trunnion to the body on the front passenger's side.
9   Remove the two bolts which secure the rack housing to the trunnion on the driver's side.
10 Lower the rack assembly, support it and then withdraw it from the driver's side of the car. Detach the trunnion end housing.
11 Overhaul of the power-assisted steering gear is not recommended, but should wear or a fault develop, change the assembly for a new or factory rebuilt unit.

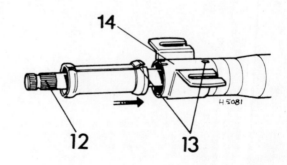

Fig. 11.27 Steering column lower bush (Sec 21)

12 Steering shaft   14 Retaining tag
13 Bush projecting tag and
    column slot

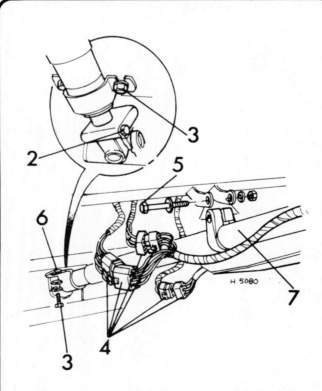

Fig. 11.26 Steering column connections (Sec 21)

2   Pinch-bolt   5   Column clamp bolt
3   Lower bracket screw   6   Lower bracket
4   Combination switch   7   Steering column
    connector plugs

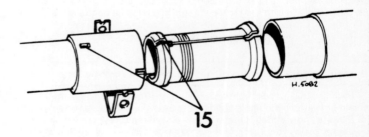

Fig. 11.28 Steering column upper bush (Sec 21)

15 Bush projecting tag and column slot

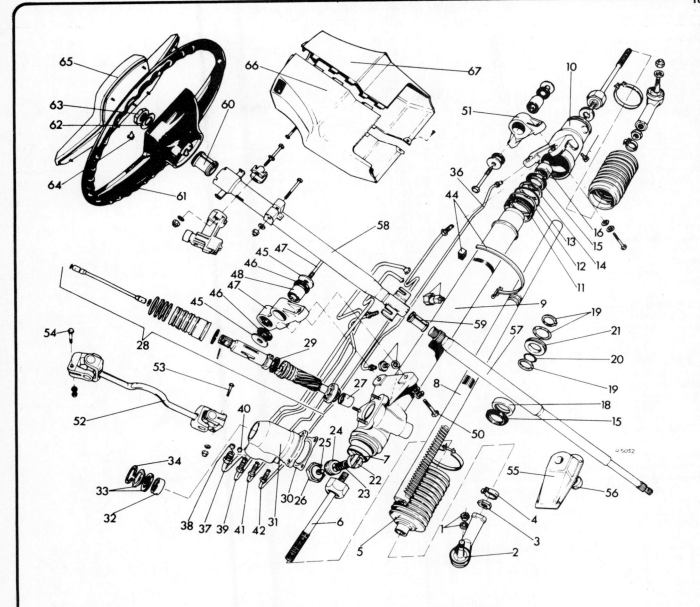

Fig. 11.29 Power-assisted steering components (Sec 22)

1	Nut and washer	18 Sleeve	35 Adaptor and seal	52 Intermediate shaft
2	Track-rod end	19 Shim circlip	36 Pipe (interconnecting)	53 Pinch-bolt
3	Locknut	20 O-ring	37 Pipe (oil return)	54 Pinch-bolt
4	Clip	21 Piston/ring assembly	38 Outlet port insert	55 Cover plate
5	Bellows	22 Rack damper plunger	39 Pipe (oil feed)	56 Seal
6	Track-rod	23 Spring	40 Insert (inlet port)	57 Steering shaft
7	Locking washer	24 Plug	41 Pipe to rack housing	58 Steering column
8	Rack	25 Screw	42 Pipe to end housing	59 Bottom bush
9	Rack housing	26 Locknut	43 Adaptor	60 Top bush
10	End housing	27 Needle bearing	44 Clip and insulator	61 Steering wheel
11	End housing nut	28 Valve/pinion assembly	45 Plain washer	62 Lockwasher
12	Spring ring	29 Valve seal	46 Buffer washer	63 Nut
13	O-ring	30 Gasket	47 Trunnion	64 Clip
14	Shim	31 Valve housing	48 Bush	65 Crash pad
15	Seal	32 Needle bearing	49 Bolt	66 Cowl
16	Bush	33 Seal and washer	50 Bolt	67 Cowl
17	Locating peg	34 Circlip	51 Trunnion	

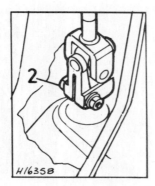

Fig. 11.30 Steering intermediate shaft pinch-bolt (2) (Sec 22)

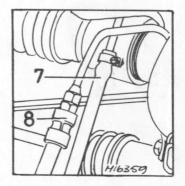

Fig. 11.31 Steering gear hoses (Sec 22)

7   Rack return hose   8   Rack feed hose

12 Before refitting, check that the protective washer is in position on top of the pinion housing. Also check that the seal is either on the protective washer or stuck in place on the body cover plate.
13 With the help of an assistant, locate the steering gear against its trunnion mountings and screw in the bolts two or three turns only.
14 Engage the intermediate shaft on the pinion so that the pinch-bolt hole and groove are aligned.
15 Tighten the mounting bolts at the pinion housing and then those at the opposite end. Reconnect the tie-rod ends.
16 Tighten the lower arm bush locknut after the weight of the car is again on its roadwheels.
17 Refill and bleed the system (Section 25).

23 Power steering fluid pump – removal and refitting

1   Cover the alternator with a plastic bag to protect it from oil spillage.
2   Hold the pump feed pipe union in one spanner to prevent it from rotating, then disconnect the feed pipe and plug it. Catch the oil which is spilled.
3   Slacken the steering pump pivot bolt and adjusting link screws.
4   Press the pump in towards the engine and remove the drivebelt.
5   Slacken the clip on the fluid return hose then disconnect it from the pump adaptor.
6   Remove the three screws which secure the pump to its mounting bracket and remove the pump (photo).
7   Refitting is a reversal of removal. Tension the drivebelt as described in Chapter 2 Section 10.
8   Bleed the system as described in Section 25.

24 Power steering fluid return hose – removal and refitting

1   Slacken the clip and disconnect the hose from the rack pipe. Catch the oil in a container.
2   Release the pipe spacing clip.
·3   Release the feed pipe clip from its stud on the transmission casing.
4   Release the return hose clip from the engine mounting bracket.
5   Cover the alternator with a plastic bag to protect it from oil spillage. Disconnect the clip and pull the return hose from the power steering fluid reservoir.
6   Refitting is a reversal of removal. Bleed the system as described in the next Section.

25 Power steering system – bleeding

1   Check that the pump reservoir is half filled with engine oil, viscosity 15W/50, to API SE/SF (Duckhams Hypergrade) up to vehicle identification number VIN 122690, or automatic transmission fluid to M2C 33G or M2C 33F (Duckhams Q-Matic) from VIN 122691.

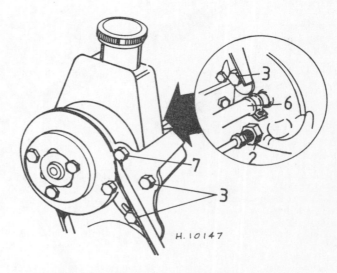

Fig. 11.32 Power steering pump (Sec 23)

2   Feed pipe union   6   Return hose
3   Mounting and adjusting bolts   7   Mounting bolt

23.6 Power steering pump bracket

2   Disconnect the LT lead which runs to the contact breaker (white/black) from the terminal on the coil.

3   Operate the starter motor for five seconds, then top up the reservoir again to the half full level. Reconnect the LT lead.

4   Centralise the steering, start the engine and let it idle.

5   Open the throttle slightly and watch the reservoir oil level. Top up immediately if it falls.

6   Drive the car forwards or backwards just enough to be able to run the steering to the left and then to the right, but without attaining full lock in either case.

7   Again centralise the steering, switch off the engine and top up the reservoir level to between the end of the dipstick and the Full Cold mark.

26 Front wheel alignment and steering angles

1   Accurate front wheel alignment is essential for good steering and slow tyre wear. Before considering the steering angle, check that the tyres are correctly inflated, that the front wheels are not buckled, the hub bearings are not worn or incorrectly adjusted and that the steering linkage is in good order, without slackness or wear at the joints.

2   Wheel alignment consists of four factors:

Camber, is the angle at which the front wheels are set from the vertical when viewed from the front of the car. Positive camber is the amount (in degrees) that the wheels are tilted outwards at the top from the vertical.

Castor, is the angle between the steering axis and a vertical line when viewed from each side of the car. Positive castor is when the steering axis is inclined rearward.

Steering axis inclination is the angle, when viewed from the front of the car, between the vertical and an imaginary line drawn between the upper and lower suspension arm swivels.

Front wheel tracking. This normally gives the front roadwheels a toe-in or toe-out but in the case of cars covered by this manual, the front roadwheels should be set parallel, or very near to it (see Specifications).

3   All steering angles, other than front wheel alignment, are set in production and cannot be altered.

4   To check front wheel alignment, place the car on level ground with the tyres correctly inflated and the front roadwheels in the straight-ahead position.

5   Remove the plug from the plunger screw of the rack housing, insert a dowel and move the steering until the dowel engages in the hole in the rack. The steering is now centralised.

6   If the tie-rod ends have been removed, set them at a datum point by releasing their locknuts and the bellows outer clips and turning each tie-rod until the roadwheels are parallel. This can best be checked by laying a length of steel rod or wood along the side of the car. When it touches all four sidewalls of the front and rear tyres, then the front wheels will be approximately parallel with each other.

7   Obtain or make a tracking gauge. One may be easily made from tubing, cranked to clear the sump and bellhousing, having an adjustable nut and setscrew at one end (photo).

26.7 Checking front wheel alignment

8   Using the gauge, measure the distance between the two inner wheel rims at hub height at the rear of the wheels.

9   Rotate the wheels (by pushing the car forwards) through 180° (half a turn) and again using the gauge, measure the distance at hub height between the two inner wheel rims at the front of the roadwheels.

10 The two measurements should be the same to give a parallel characteristic.

11 Where this is not the case, turn each track-rod an equal amount and when adjustment is correct and has been re-checked, tighten the track-rod end locknuts making sure that the track-rod ends are in their correct attitude (centre of their arcs of travel).

12 Check that the steering rack bellows are not twisted and tighten the securing clips.

13 Remove the gauge and rack dowel pin and refit the rack plug.

27 Rear wheel alignment

1   The rear wheel alignment and suspension angles are set in production and cannot be adjusted.

2   Any deviation from the tolerances given in Specifications can only be due to severely worn suspension bushes, loose mountings or collision damage which must be rectified immediately.

Fault diagnosis appears overleaf

28 Fault diagnosis – suspension and steering

Symptom	Reason(s)
Steering feels vague, car 'wanders' and 'floats' at speed	General wear or damage Tyre pressures uneven Steering gear or suspension balljoints badly worn Suspension geometry incorrect Steering mechanism free play excessive Front suspension and rear suspension pick-up points out of alignment
Stiff and heavy steering	Lack of maintenance or accident damage Tyre pressures too low Front wheel alignment incorrect Suspension geometry incorrect Steering gear incorrectly adjusted too tightly Steering column badly misaligned
Wheel wobble and vibration	General wear or damage Wheel nuts loose Front wheels and tyres out of balance Steering or suspension balljoints badly worn Hub bearings badly worn Steering gear free play excessive

Chapter 12 Bodywork

Contents

1 General description

The bodywork is of all-steel welded construction. The car is produced in five-door style only and most essential accessories are included as standard. Both front seats are of reclining type and the driver's seat is also adjustable for height. A floor console is fitted and the rear seat has a centre armrest.

In order to prevent distortion of the body, it is most important that the jacking and towing procedures given at the front of this manual are strictly followed.

The spare wheel is stowed on the left-hand side of the luggage boot, and the tools are located under the flap in the luggage boot floor.

2 Maintenance – bodywork and underframe

1 The general condition of a vehicle's bodywork is the one thing that significantly affects its value. Maintenance is easy but needs to be regular. Neglect, particularly after minor damage, can lead quickly to further deterioration and costly repair bills. It is important also to keep watch on those parts of the vehicle not immediately visible, for instance the underside, inside all the wheel arches and the lower part of the engine compartment.

2 The basic maintenance routine for the bodywork is washing – preferably with a lot of water, from a hose. This will remove all the loose solids which may have stuck to the vehicle. It is important to flush these off in such a way as to prevent grit from scratching the finish. The wheel arches and underframe need washing in the same way to remove any accumulated mud which will retain moisture and tend to encourage rust. Paradoxically enough, the best time to clean the underframe and wheel arches is in wet weather when the mud is thoroughly wet and soft. In very wet weather the underframe is usually cleaned of large accumulations automatically and this is a good time for inspection.

3 Periodically, it is a good idea to have the whole of the underframe of the vehicle steam cleaned, engine compartment included, so that a thorough inspection can be carried out to see what minor repairs and renovations are necessary. Steam cleaning is available at many garages and is necessary for removal of the accumulation of oily grime which sometimes is allowed to become thick in certain areas near the engine and transmission. If steam cleaning facilities are not available, there are one or two excellent grease solvents available which can be brush applied. The dirt can then be simply hosed off.

4 After washing paintwork, wipe off with a chamois leather to give an unspotted clear finish. A coat of clear protective wax polish will give added protection against chemical pollutants in the air. If the paintwork sheen has dulled or oxidised, use a cleaner/polisher combination to restore the brilliance of the shine. This requires a little effort, but such dulling is usually caused because regular washing has been neglected. Always check that the door body sill and ventilator opening drain holes and pipes are completely clear so that water can be drained out. Bright work should be treated in the same way as paintwork. Windscreens and windows can be kept clear of the smeary film which often appears, by adding a little ammonia to the water. If they are scratched, a good rub with a proprietary metal polish will often clear them. Never use any form of wax or other body or chromium polish on glass (photos).

2.4A Door drain hole

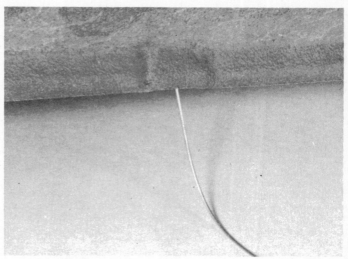

2.4B Sill drain hole

3   Maintenance – upholstery and carpets

1   Mats and carpets should be brushed or vacuum cleaned regularly to keep them free of grit. If they are badly stained remove them from the vehicle for scrubbing or sponging and make quite sure they are dry before refitting. Seats and interior trim panels can be kept clean by wiping with a damp cloth. If they do become stained (which can be more apparent on light coloured upholstery) use a little liquid detergent and a soft nail brush to scour the grime out of the grain of the material. Do not forget to keep the headlining clean in the same way as the upholstery. When using liquid cleaners inside the vehicle do not over-wet the surfaces being cleaned. Excessive damp could get into the seams and padded interior causing stains, offensive odours or even rot. If the inside of the vehicle gets wet accidentally it is worthwhile taking some trouble to dry it out properly, particularly where carpets are involved. *Do not leave oil or electric heaters inside the vehicle for this purpose.*

4   Minor body damage – repair

The photographic sequences on pages 198 and 199 illustrate the operations detailed in the following sub-sections.

Repair of minor scratches in bodywork
If the scratch is very superficial, and does not penetrate to the metal of the bodywork, repair is very simple. Lightly rub the area of the scratch with a paintwork renovator, or a very fine cutting paste, to remove loose paint from the scratch and to clear the surrounding bodywork of wax polish. Rinse the area with clean water.
Apply touch-up paint to the scratch using a fine paint brush; continue to apply fine layers of paint until the surface of the paint in the scratch is level with the surrounding paintwork. Allow the new paint at least two weeks to harden; then blend it into the surrounding paintwork by rubbing the scratch area with a paintwork renovator or a very fine cutting paste. Finally, apply wax polish.
Where the scratch has penetrated right through to the metal of the bodywork, causing the metal to rust, a different repair technique is required. Remove any loose rust from the bottom of the scratch with a penknife, then apply rust inhibiting paint to prevent the formation of rust in the future. Using a rubber or nylon applicator fill the scratch with bodystopper paste. If required, this paste can be mixed with cellulose thinners to provide a very thin paste which is ideal for filling narrow scratches. Before the stopper-paste in the scratch hardens, wrap a piece of smooth cotton rag around the top of a finger. Dip the finger in cellulose thinners and then quickly sweep it across the surface of the stopper-paste in the scratch; this will ensure that the surface of

the stopper-paste is slightly hollowed. The scratch can now be painted over as described earlier in this Section.

Repair of dents in bodywork
When deep denting of the vehicle's bodywork has taken place, the first task is to pull the dent out, until the affected bodywork almost attains its original shape. There is little point in trying to restore the original shape completely, as the metal in the damaged area will have stretched on impact and cannot be reshaped fully to its original contour. It is better to bring the level of the dent up to a point which is about $\frac{1}{8}$ in (3 mm) below the level of the surrounding bodywork. In cases where the dent is very shallow anyway, it is not worth trying to pull it out at all. If the underside of the dent is accessible, it can be hammered out gently from behind, using a mallet with a wooden or plastic head. Whilst doing this, hold a suitable block of wood firmly against the outside of the panel to absorb the impact from the hammer blows and thus prevent a large area of the bodywork from being 'belled-out'.
Should the dent be in a section of the bodywork which has a double skin or some other factor making it inaccessible from behind, a different technique is called for. Drill several small holes through the metal inside the area – particularly in the deeper section. Then screw long self-tapping screws into the holes just sufficiently for them to gain a good purchase in the metal. Now the dent can be pulled out by pulling on the protruding heads of the screws with a pair of pliers.
The next stage of the repair is the removal of the paint from the damaged area, and from an inch or so of the surrounding 'sound' bodywork. This is accomplished most easily by using a wire brush or abrasive pad on a power drill, although it can be done just as effectively by hand using sheets of abrasive paper. To complete the preparation for filling, score the surface of the bare metal with a screwdriver or the tang of a file, or alternatively, drill small holes in the affected area. This will provide a really good 'key' for the filler paste.
To complete the repair see the Section on filling and re-spraying.

Bodywork repairs – filling and re-spraying
Before using this Section, see the Sections on dent, deep scratch, rust holes and gash repairs.
Many types of bodyfiller are available, but generally speaking those proprietary kits which contain a tin of filler paste and a tube of resin hardener are best for this type of repair. A wide, flexible plastic or nylon applicator will be found invaluable for imparting a smooth and well contoured finish to the surface of the filler.
Mix up a little filler on a clean piece of card or board – measure the hardener carefully (follow the maker's instructions on the pack) otherwise the filler will set too rapidly or too slowly.
Using the applicator apply the filler paste to the prepared area; draw the applicator across the surface of the filler to achieve the

correct contour and to level the filler surface. As soon as a contour that approximates to the correct one is achieved, stop working the paste — if you carry on too long the paste will become sticky and begin to 'pick up' on the applicator. Continue to add thin layers of filler paste at twenty-minute intervals until the level of the filler is just proud of the surrounding bodywork.

Once the filler has hardened, excess can be removed using a metal plane or file. From then on, progressively finer grades of abrasive paper should be used, starting with a 40 grade production paper and finishing with 400 grade wet-and-dry paper. Always wrap the abrasive paper around a flat rubber, cork, or wooden block — otherwise the surface of the filler will not be completely flat. During the smoothing of the filler surface the wet-and-dry paper should be periodically rinsed in water. This will ensure that a very smooth finish is imparted to the filler at the final stage.

At this stage the 'dent' should be surrounded by a ring of bare metal, which in turn should be encircled by the finely 'feathered' edge of the good paintwork. Rinse the repair area with clean water, until all of the dust produced by the rubbing-down operation has gone.

Spray the whole repair area with a light coat of primer — this will show up any imperfections in the surface of the filler. Repair these imperfections with fresh filler paste or bodystopper, and once more smooth the surface with abrasive paper. If bodystopper is used, it can be mixed with cellulose thinners to form a really thin paste which is ideal for filling small holes. Repeat this spray and repair procedure until you are satisfied that the surface of the filler, and the feathered edge of the paintwork are perfect. Clean the repair area with clean water and allow to dry fully.

The repair area is now ready for final spraying. Paint spraying must be carried out in a warm, dry, windless and dust free atmosphere. This condition can be created artificially if you have access to a large indoor working area, but if you are forced to work in the open, you will have to pick your day very carefully. If you are working indoors, dousing the floor in the work area with water will help to settle the dust which would otherwise be in the atmosphere. If the repair area is confined to one body panel, mask off the surrounding panels; this will help to minimise the effects of a slight mis-match in paint colours. Bodywork fittings (eg chrome strips, door handles etc) will also need to be masked off. Use genuine masking tape and several thicknesses of newspaper for the masking operations.

Before commencing to spray, agitate the aerosol can thoroughly, then spray a test area (an old tin, or similar) until the technique is mastered. Cover the repair area with a thick coat of primer; the thickness should be built up using several thin layers of paint rather than one thick one. Using 400 grade wet-and-dry paper, rub down the surface of the primer until it is really smooth. While doing this, the work area should be thoroughly doused with water, and the wet-and-dry paper periodically rinsed in water. Allow to dry before spraying on more paint.

Spray on the top coat, again building up the thickness by using several thin layers of paint. Start spraying in the centre of the repair area and then, using a circular motion, work outwards until the whole repair area and about 2 inches of the surrounding original paintwork is covered. Remove all masking material 10 to 15 minutes after spraying on the final coat of paint.

Allow the new paint at least two weeks to harden, then, using a paintwork renovator or a very fine cutting paste, blend the edges of the paint into the existing paintwork. Finally, apply wax polish.

Repair of rust holes or gashes in bodywork

Remove all paint from the affected area and from an inch or so of the surrounding 'sound' bodywork, using an abrasive pad or a wire brush on a power drill. If these are not available a few sheets of abrasive paper will do the job just as effectively. With the paint removed you will be able to gauge the severity of the corrosion and therefore decide whether to renew the whole panel (if this is possible) or to repair the affected area. New body panels are not as expensive as most people think and it is often quicker and more satisfactory to fit a new panel than to attempt to repair large areas of corrosion.

Remove all fittings from the affected area except those which will act as a guide to the original shape of the damaged bodywork (eg headlamp shells etc). Then, using tin snips or a hacksaw blade, remove all loose metal and any other metal badly affected by corrosion. Hammer the edges of the hole inwards in order to create a slight depression for the filler paste.

Wire brush the affected area to remove the powdery rust from the surface of the remaining metal. Paint the affected area with rust inhibiting paint; if the back of the rusted area is accessible treat this also.

Before filling can take place it will be necessary to block the hole in some way. This can be achieved by the use of zinc gauze or aluminium tape.

Zinc gauze is probably the best material to use for a large hole. Cut a piece to the approximate size and shape of the hole to be filled, then position it in the hole so that its edges are below the level of the surrounding bodywork. It can be retained in position by several blobs of filler paste around its periphery.

Aluminium tape should be used for small or very narrow holes. Pull a piece off the roll and trim it to the approximate size and shape required, then pull off the backing paper (if used) and stick the tape over the hole; it can be overlapped if the thickness of one piece is insufficient. Burnish down the edges of the tape with the handle of a screwdriver or similar, to ensure that the tape is securely attached to the metal underneath.

5 Major body damage — repair

Where serious damage has occurred or large areas need renewal due to neglect, it means certainly that completely new sections or panels will need welding in and this is best left to professionals. If the damage is due to impact it will also be necessary to completely check the alignment of the body shell structure. Due to the principle of construction the strength and shape of the whole car can be affected by damage to a part. In such instances the services of a workshop with specialist checking jigs are essential. If a body is left misaligned it is first of all dangerous as the car will not handle properly and secondly uneven stresses will be imposed on the steering, engine and transmission, causing abnormal wear or complete failure. Tyre wear may also be excessive.

6 Maintenance — hinges and locks

1 Oil the hinges of the bonnet, boot and doors with a drop or two or light oil periodically. A good time is after the car has been washed.
2 Oil the bonnet release, the catch pivot pin and the safety catch pivot pin periodically.
3 Do not over lubricate door latches and strikers. Normally a little oil on the rotary cam spindle alone is sufficient.

7 Radiator grille — removal and refitting

1 Open the bonnet.
2 Remove the four screws from the top edge of the grille and the three screws from the bottom (photo).

7.2 Radiator grille lower screw

7.3 Removing radiator grille

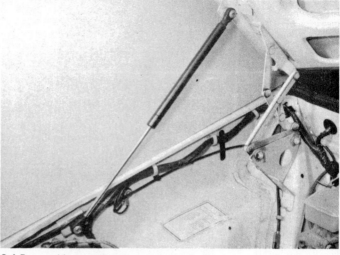

8.4 Bonnet hinge and strut

3   Withdraw the grille from the car (photo).
4   Refitting is a reversal of removal.

8   Bonnet – removal and refitting

1   Open the bonnet and support it in the fully open position.
2   Remove the bonnet lock striker (two bolts).
3   Remove the sound insulating pad after extracting the securing clips.
4   Have an assistant support the bonnet and then unscrew the bolts which attach the support struts to the bonnet (photo).
5   If an engine compartment lamp is fitted, the leads must be disconnected at the snap-connectors.
6   Mark the position of the hinge plates on the underside of the bonnet and then remove the hinge securing screws. Lift the bonnet from the car.
7   Refitting is a reversal of removal.

9   Bonnet lock – removal and refitting

1   Release the radiator from the bonnet lock platform (two bolts) to provide access to the lock.
2   Disconnect the lock remote control cable from the lock lever.
3   Unbolt the lock assembly and withdraw it downwards from the lock platform.
4   If the cable is to be removed, disconnect the outer cable from its clip on the bonnet lock platform and draw the cable through its three plastic clips.
5   Disconnect the inner cable from the bonnet release hand control (one bolt) and detach the outer cable clip from the hand control mounting bracket.
6   Withdraw the cable assembly into the interior of the car.
7   Refitting is a reversal of removal but before tightening the lock bolts, close the bonnet to centralise the lock in relation to the striker. The latter can be adjusted to ensure correct bonnet closure if the locknut is released and the centre screw turned.

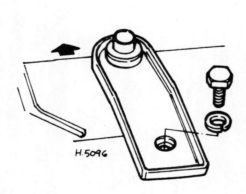

Fig. 12.1 Radiator top support bracket (Sec 9)

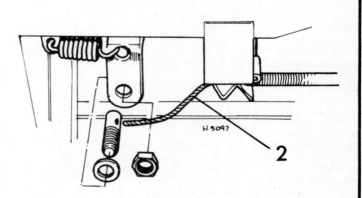

Fig. 12.2 Bonnet lock cable (2) and associated components
(Sec 9)

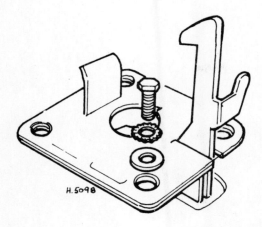

Fig. 12.3 Bonnet lock and safety catch (Sec 9)

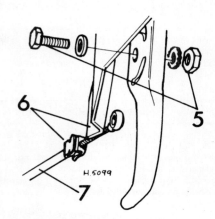

Fig. 12.4 Bonnet release control handle (Sec 9)

5   Cable bolt and nut   7   Outer cable
6   Outer cable clip and bracket

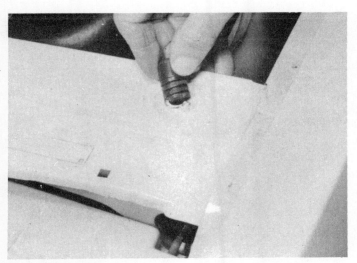

9.8 Bonnet buffer

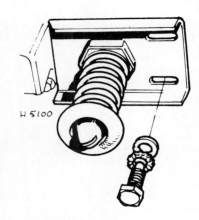

Fig. 12.5 Bonnet lock dovetail striker (Sec 9)

8   The height of the bonnet (closed) can also be adjusted to stop
rattles by screwing the rubber buffer stops in or out (photo).

10 Bumpers – removal and refitting

1   The bumpers are mounted on brackets which in turn are bolted to
the underframe members.
2   The bolt heads are recessed in the bumper and covered with
blanking caps.
3   Unscrew the nuts from inside the bumper and lift the bumper from
the brackets.
4   Before the rear bumper can be fully removed from the car, the
leads to the number plate lamps must be disconnected.
5   Refitting is a reversal of removal.

11 Front wheel arch liner – removal and refitting

1   In the interest of body protection, liners are fitted under the front
wings. Unless they are removed at reasonably regular intervals and the
undersurface of the wing given a coat of protective paint, these

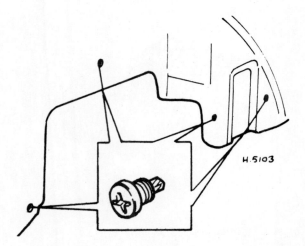

Fig. 12.6 Front wheel arch liner screws (Sec 11)

components may, in fact, have the opposite effect and actually increase the rate of corrosion due to the build up of damp behind them.

2   Jack-up the front of the car, support it securely and remove the roadwheels.

3   Extract the screws which secure the liners to the valance.

4   Pull the front of the liner from the wing and then using a flat lever prise it from the turned-over edge of the wing.

5   Refitting is a reversal of removal but insert the rear of the liner first under the wheel arch.

12 Door trim panel – removal and refitting

1   Extract the two screws from the lower face of the door armrest (photo).

2   Swing the armrest into the vertical position and pull it from the door. The armrest has a bayonet type fixing.

3   Unscrew and remove the door locking plunger knob (photo).

4   Extract the fixing screw and withdraw the window regulator handle (photos).

5   Extract the screw and withdraw the bezel from the door lock remote control. Do not remove the door glass demister escutcheon.

6   Insert a broad blade between the bottom edge of the trim panel and the door. Prise the trim panel free, disengaging the fixing clips.

7   Prise the rest of the trim panel free using the fingers (photo).

12.1 Door armrest

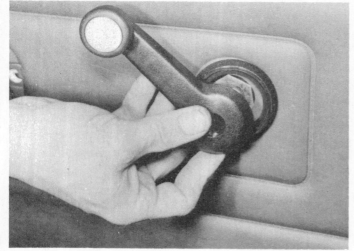

12.3 Door lock plunger

12.4A Window regulator screw

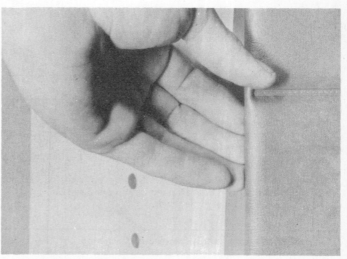

12.4B Removing window regulator handle

12.7 Removing door trim panel

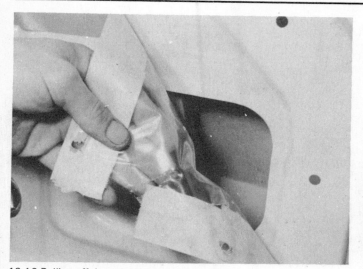
12.10 Pulling off door weatherproof sheet

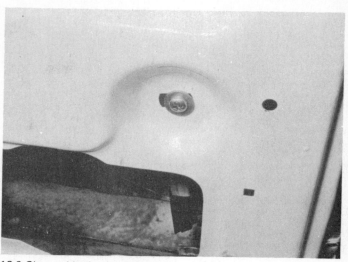
13.2 Glass guide channel screw

8   Pull the trim panel just far enough away from the door to be able to reach behind it and disconnect the speaker wires.
9   Remove the trim panel. The radio speaker can be removed if necessary after drilling out the rivets.
10 Detach the waterproof sheet from the internal face of the door (photo).
11 Refitting is a reversal of removal.

13 Front door lock — removal and refitting

1   Wind the glass fully upward and then remove the door trim panel as described in the preceding Section.
2   Remove the screw which retains the glass rear channel to the door panel. Move the lower end of the channel forward to provide access to the lock linkage (photo).
3   Remove the screw which secures the remote control to the door inner panel. Slide the control to the rear to release it from the door panel (photo).
4   Unclip and remove the remote control link from the door and lock (photo).
5   Unclip and remove the door lock link from the lock cylinder assembly.

13.3A Door remote control handle

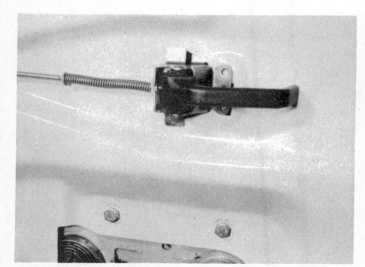
13.3B Remote control handle, door trim panel removed

13.4A Door cylinder lock retainer

13.4B Door exterior handle fixing nuts

13.7 Door lock

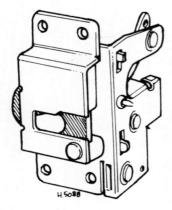

Fig. 12.7 Door lock (Sec 13)

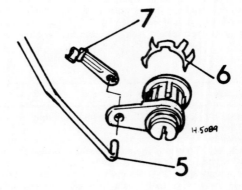

Fig. 12.8 Door lock cylinder (Sec 13)

5   Link   7   Link clip
6   Lock retainer

6   Disconnect the nylon connecting link from the ball end of the door exterior handle.
7   Remove the four screws which hold the lock assembly to the door and with the lock in its locked position, withdraw it into the door cavity, then remove it through the large aperture in the door interior panel (photo).
8   Unclip and detach the lock cylinder assembly and the sill plunger links from the lock.
9   Refitting is a reversal of removal but adjust the nylon connecting link attached to the ball end of the door exterior handle so that with the lock in the unlocked state, the lower end of the locking lever is just clear of the lock trigger plate when the nylon link is aligned to the ball end.
10 If the speaker was removed, use self-tapping screws and spire nuts to refix it to the trim panel.
11 Make quite sure that the coloured wire is connected to the matching coloured terminal on the speaker.
12 On cars equipped with a central door locking system, refer to Chapter 10, Section 30.

14 Front door window regulator – removal and refitting

1   Raise the glass to the fully closed position.
2   Remove the door interior panel, as described in Section 12.
3   Remove the waterproof sheet from the internal face of the door.

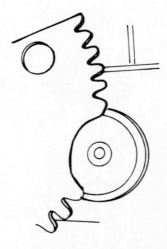

Fig. 12.9 Window regulator setting prior to fitting (Sec 14)

Six teeth above pinion

4   Wedge the glass in the fully closed position by using two pieces of wood.

5   Remove the four screws which secure the regulator unit to the door panel, push the regulator inwards and forwards to disconnect it from the slides of the glass lifting channels.

6   Withdraw the regulator through the large aperture in the door inner panel.

7   Commence refitting by winding the regulator until six teeth are visible above the pinion housing (fully raised position) and then reverse the removal operations.

8   Engage the regulator arms in the slides in the following order:

    (a)   *door inner panel slide*
    (b)   *glass channel rear slide*
    (c)   *glass channel front slide*

9   Prior to fully tightening the regulator unit screws, wind the window down slightly and adjust the position of the glass within its guides.

10 On cars equipped with electrically-operated windows, refer to Chapter 10, Section 29.

15 Front door glass – removal and refitting

1   Wind the glass down to its lowest position, remove the door interior trim panel, as described in Section 12, and remove the waterproof sheet.

2   Pull the glass outer sealing strip upwards to release it from the clips of the sill.

3   Remove the two screws and release the lower ends of the door glass guides from the door inner panel.

4   Position the glass inside the door so that the guides are between the glass and the door outer panel.

5   Push the glass forward and disconnect the glass lifting channel from the front arm of the regulator then move it to the rear and disconnect the channel from the rear arm of the regulator.

6   Withdraw the glass upwards to the outer side of the door, at the same time tilting it as shown in Fig. 12.11.

7   The channels can be driven from the glass using a hammer and a small block of hardwood.

8   Refitting is a reversal of removal but position the lifting channel as indicated in the diagram (Fig. 12.12).

16 Front door – removal and refitting

1   Remove the door internal trim panel, as described in Section 12.

2   Open the door fully and support its lower edge on jacks or blocks with pads of rag to prevent damage to the paintwork. Mark the position of the hinge plates on the door.

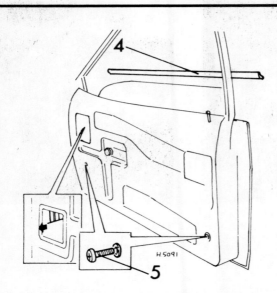

Fig. 12.10 Front door glass guides and fixing screws (Sec 15)

    4   Sealing strip   *5   Lower fixing screws*

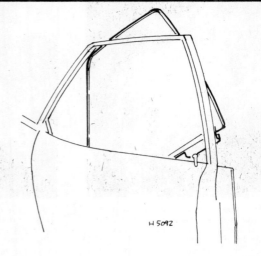

Fig. 12.11 Removing front door glass (Sec 15)

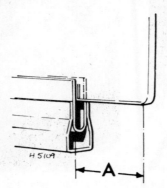

Fig. 12.12 Front door glass channel fitting diagram (Sec 15)

    A   Measured from rear edge of glass 1.5 in (38.0 mm)

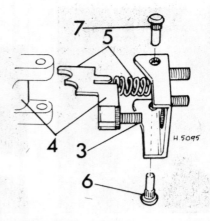

Fig. 12.13 Door lower hinge components (Sec 16)

  3   Check stud   *6   Hinge lower pin*
  4   Check plate groove and roller   *7   Hinge upper pin*
  5   Check plate and spring

1

This photographic sequence shows the steps taken to repair the dent and paintwork damage shown above. In general, the procedure for repairing a hole will be similar; where there are substantial differences, the procedure is clearly described and shown in a separate photograph.

2

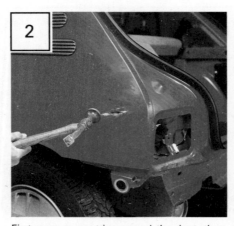

First remove any trim around the dent, then hammer out the dent where access is possible. This will minimise filling. Here, after the large dent has been hammered out, the damaged area is being made slightly concave.

3

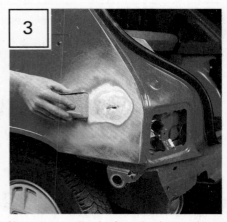

Next, remove all paint from the damaged area by rubbing with coarse abrasive paper or using a power drill fitted with a wire brush or abrasive pad. 'Feather' the edge of the boundary with good paintwork using a finer grade of abrasive paper.

4

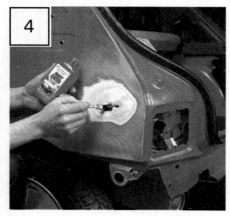

Where there are holes or other damage, the sheet metal should be cut away before proceeding further. The damaged area and any signs of rust should be treated with Turtle Wax Hi-Tech Rust Eater, which will also inhibit further rust formation.

5

For a large dent or hole mix Holts Body Plus Resin and Hardener according to the manufacturer's instructions and apply around the edge of the repair. Press Glass Fibre Matting over the repair area and leave for 20-30 minutes to harden. Then ...

5A

... brush more Holts Body Plus Resin and Hardener onto the matting and leave to harden. Repeat the sequence with two or three layers of matting, checking that the final layer is lower than the surrounding area. Apply Holts Body Plus Filler Paste as shown in Step 5B.

5B

For a medium dent, mix Holts Body Plus Filler Paste and Hardener according to the manufacturer's instructions and apply it with a flexible applicator. Apply thin layers of filler at 20-minute intervals, until the filler surface is slightly proud of the surrounding bodywork.

5C

For small dents and scratches use Holts No Mix Filler Paste straight from the tube. Apply it according to the instructions in thin layers, using the spatula provided. It will harden in minutes if applied outdoors and may then be used as its own knifing putty.

6

Use a plane or file for initial shaping. Then, using progressively finer grades of wet-and-dry paper, wrapped round a sanding block, and copious amounts of clean water, rub down the filler until glass smooth. 'Feather' the edges of adjoining paintwork.

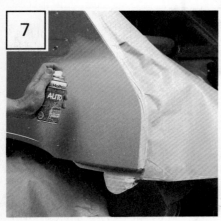

7

Protect adjoining areas before spraying the whole repair area and at least one inch of the surrounding sound paintwork with Holts Dupli-Color primer.

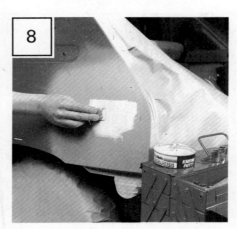

8

Fill any imperfections in the filler surface with a small amount of Holts Body Plus Knifing Putty. Using plenty of clean water, rub down the surface with a fine grade wet-and-dry paper – 400 grade is recommended – until it is really smooth.

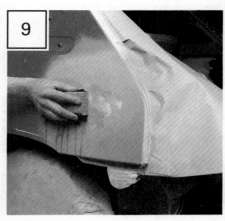

9

Carefully fill any remaining imperfections with knifing putty before applying the last coat of primer. Then rub down the surface with Holts Body Plus Rubbing Compound to ensure a really smooth surface.

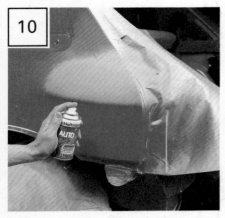

10

Protect surrounding areas from overspray before applying the topcoat in several thin layers. Agitate Holts Dupli-Color aerosol thoroughly. Start at the repair centre, spraying outwards with a side-to-side motion.

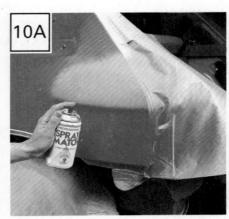

10A

If the exact colour is not available off the shelf, local Holts Professional Spraymatch Centres will custom fill an aerosol to match perfectly.

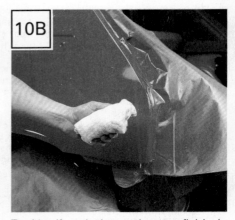

10B

To identify whether a lacquer finish is required, rub a painted unrepaired part of the body with wax and a clean cloth.

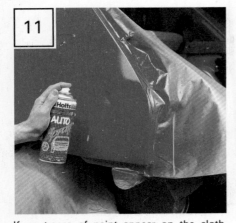

11

If *no* traces of paint appear on the cloth, spray Holts Dupli-Color clear lacquer over the repaired area to achieve the correct gloss level.

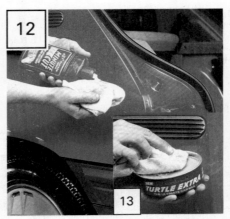

12

13

The paint will take about two weeks to harden fully. After this time it can be 'cut' with a mild cutting compound such as Turtle Wax Minute Cut prior to polishing with a final coating of Turtle Wax Extra.

14

When carrying out bodywork repairs, remember that the quality of the finished job is proportional to the time and effort expended.

16.4A Door upper hinge

16.4B Door lower hinge

16.11 Door lock striker

3   Remove the six nuts and two plates and lift the door from the hinge plates.

4   If the hinges are to be removed, mark their position on the body pillars (photos).

5   To remove the lower hinge, protect the threads on the door check stud, support the hinge plate and drive the stud from it.

6   Fully open the hinge and drive the door check plate forwards so that its second detent groove engages with the detent roller. Pull the checkplate and spring from the hinge.

7   Drive the hinge lower pin downwards and remove it.

8   Drive the hinge upper pin upwards and remove it.

9   The fixed plates of the hinges are secured to the body by welding and if worn or distorted they will have to be renewed by your dealer or a body repair shop.

10   Reassembly and refitting are reversals of dismantling and removal but align the doors within and flush with, the body shell before finally tightening the hinge nuts.

11   Door closure can be adjusted by loosening the striker on the pillar and moving it slightly (photo).

17 Rear door lock – removal and refitting

This is very similar to the procedure described in Section 13 except, of course, that a lock cylinder is not incorporated in the system.

18 Rear door window regulator – removal and refitting

1   Wind the window glass fully down.

2   Remove the door interior trim panel, as described in Section 12.

3   Remove the waterproof sheet from the internal face of the door.

4   Support the glass and remove the four screws which retain the regulator unit.

5   Push the regulator unit inwards to the rear to disengage it from the glass channel.

6   Withdraw the regulator unit through the large aperture in the door panel.

7   Refitting is a reversal of removal but engage the regulator arms in the slides in the following order:

  (a) glass channel front slide
  (b) glass channel rear slide
  (c) door inner panel slide

19 Rear door glass – removal and refitting

1   The operations are similar to those described for the front door glass in the preceding Section, except that the rivet which secures the top of the glass rear guide to the top of the door must be drilled out.

2   Pull the top of the guide forward to release it from the fixed glass rubber. Rotate the guide so that its upper mounting bracket is pointing towards the door and withdraw the guide. The door sill slot will need prising open slightly to ease withdrawal.

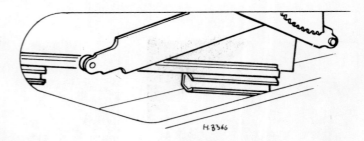

H.8366

Fig. 12.14 Window regulator disconnected from channel (Sec 19)

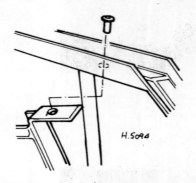

H.5094

Fig. 12.15 Rear door glass guide securing rivet (Sec 19)

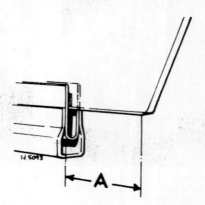

H.5073

A

Fig. 12.16 Rear door glass channel fitting diagram (Sec 19)

A   Measured from rear edge of glass 1.10 in (28.0 mm)

3   Withdraw the fixed glass and its rubber surround in a forward direction from the door.
4   Withdraw the main glass panel upwards and towards the inner side of the door after disengaging the regulator arms from the glass channels.
5   Refitting is a reversal of removal but position the lifting channel on the glass in accordance with the diagram (Fig. 12.16) and insert a new pop rivet at the top of the rear guide.

20 Rear door – removal and refitting

The procedure is similar to that described for a front door in Section 16, except that the door glass front guide must be moved aside (one screw) to gain access to the hinge nuts.

21 Tailgate – removal and refitting

1   Open the tailgate and disconnect the electrical leads from the window heater element.
2   Disconnect the leads from the tailgate wiper motor.
3   Disconnect the washer fluid hose which runs to the tailgate jet.
4   Mark round the hinges on the underside of the tailgate. Use a pencil or masking tape for this.
5   Have an assistant support the weight of the tailgate while the supporting struts are disconnected (photo).
6   Unscrew the hinge bolts from the tailgate and lift it from the car (photo).
7   Refitting is a reversal of removal, but do not fully tighten the bolts until the alignment of the tailgate within the body aperture is checked at closure. Adjust the position of the tailgate if necessary and then fully tighten the hinge bolts.

22 Tailgate lock – removal and refitting

1   Open the tailgate fully.
2   Unbolt (four screws) the lock from inside the boot lid (photo).
3   Withdraw the lock enough to be able to remove the clip which secures the lock link to the lever on the cylinder assembly. Disconnect the link from the lever.
4   To remove the lock cylinder assembly, prise open the retaining clip and release it from the shoulder on the lock body. Withdraw the cylinder lock from the outside.
5   Refitting is a reversal of removal but note that the lugs on the lock cylinder body are of different widths to ensure that it can only be fitted one way.
6   Adjust the striker to ensure positive closure (photo).

23 Windscreen and tailgate glass – removal and refitting

1   This is a job best left to the professionals, but if you must carry out the work yourself, the operations are described in the following paragraphs.
2   Where the screen is to be removed intact or is of laminated type but cracked, then an assistant will be required. First release the rubber surround from the bodywork by running a blunt, small screwdriver around and under the rubber weatherstrip both inside and outside the car. This operation will break the adhesion of the sealer originally used. Take care not to damage the paintwork or catch the rubber surround with the screwdriver. Remove the windscreen wiper arms and interior mirror and place a protective cover on the bonnet.
3   Have your assistant push the inner lip of the rubber surround off the flange of the windscreen body aperture. Commence pushing the glass at one of the upper corners. Once the rubber surround starts to

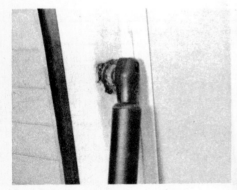
21.5 Tailgate strut upper mounting

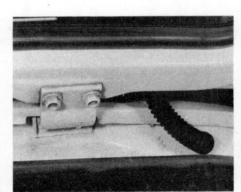
21.6 Tailgate hinge

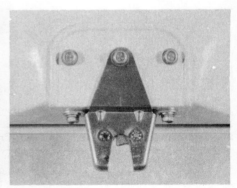
22.2 Tailgate lock

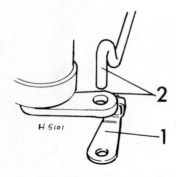

Fig. 12.17 Tailgate lock link clip (1) and link (2) (Sec 22)

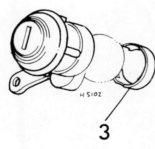

Fig. 12.18 Tailgate lock cylinder (Sec 22)

3   Retaining clip

22.6 Tailgate striker

peel off the flange, the screen may be forced gently outwards by careful hand pressure. The second person should support and remove the screen complete with rubber surround and bright trim as it comes out.

4   Remove the bright trim from the rubber surround.

5   Before fitting a windscreen, ensure that the rubber surround is completely free from old sealant and glass fragments, and has not hardened or cracked. Fit the rubber surround to the glass and apply a bead of suitable sealant between the glass outer edge and the rubber.

6   Cut a piece of strong cord greater in length that the periphery of the glass and insert it into the body flange locating channel of the rubber surround.

7   Apply a thin bead of sealant to the face of the rubber channel which will eventually mate with the body.

8   Offer up the windscreen to the body aperture and pass the ends of the cord, previously fitted and located at bottom centre, into the vehicle interior.

9   Press the windscreen into place, at the same time have an assistant pulling the cords to engage the lip of the rubber channel over the body flange.

10   Remove any excess sealant with paraffin soaked rag and fit the bright trim.

11   Removal and refitting of the rear window glass is carried out in an identical manner but first disconnect the leads to the heating element in the glass.

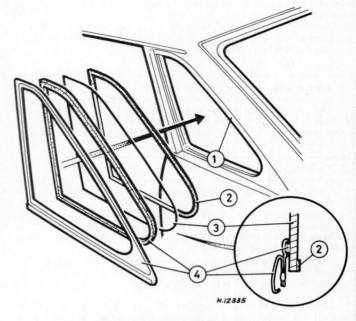

Fig. 12.19 Rear quarter-light (Sec 24)

1 Flange   4 Double sided tape and
2 Sealer strip   finisher
3 Glass

24 Rear quarter-light – removal and refitting

1   Ease the top edge of the glass away from the sealer strip, cutting the sealer with a knife if necessary to release it.

2   If the original assembly is being refitted, do not attempt to pull off the finisher.

3   Before refitting, clean away all old sealer and apply glass primer to the body flange recess and to a 10.0 mm strip around the mating surface of the glass.

25 Front seat – removal and refitting

1   Prior to removing the front seat disconnect the seat belt switch lead at the plug.

2   Remove the four screws and release the seat slides from the floor. Remove the seat (photos).

3   Refitting is a reversal of removal but note that the large flat washers are located under the carpet.

26 Rear seat squab remote control cable – removal and refitting

1   Open the tailgate and a rear door.

2   Release the squab cable from the lock trunnion.

3   Pull the inner cable from the release handle then remove the outer cable.

4   Refitting is a reversal of removal.

27 Rear seat and squab – removal and refitting

1   Open both rear doors and the tailgate.

2   Release and tilt the rear squab forward.

3   Extract the screws from the squab pivot brackets.

4   Peel back the bottom corners of the squab backing carpet, extract the hinge bracket screws and remove the squab complete from the car.

5   Extract the seat pivot bracket screws and withdraw the seat.

6   The armrest can be removed if it is partially opened and the screws which hold the bottom pivots to the running channel removed. Tilt the armrest sideways, disengage the pivots and their rollers and withdraw the armrest.

7   Refitting is a reversal of removal, but do not tighten the screws until the seat and the squab have been centralised between the door flanges.

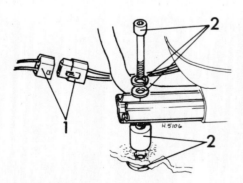

Fig. 12.20 Front seat belt switch lead connector (1), fixing bolt washers and spacer (2) (Sec 25)

25.2A Front seat front mounting

25.2B Front seat rear mounting

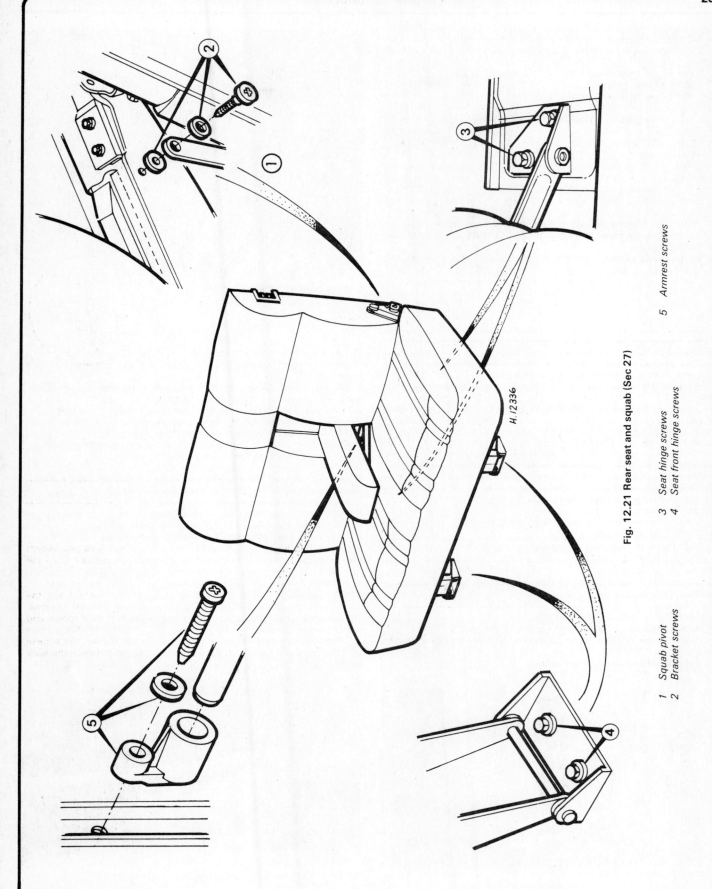

H.12336

Fig. 12.21 Rear seat and squab (Sec 27)

1 Squab pivot
2 Bracket screws

3 Seat hinge screws
4 Seat front hinge screws

5 Armrest screws

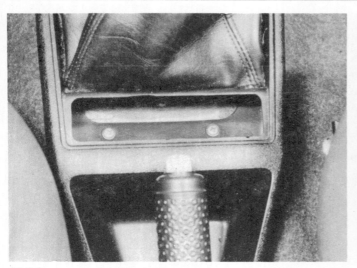

28.3 Centre console rear mounting screws

28.4 Centre console side fixing screws

28 Centre console – removal and refitting

1 Pull the flexible gaiters up the seat belt stalks.
2 Release the locknut and unscrew the gear lever or speed selector knob.
3 Prise out the blanking plate from the rear end of the console (photo).
4 Extract the two screws now exposed and the two on either side at the front end of the console (photo).
5 On cars equipped with automatic transmission, move the selector lever to position 1 or 2 and prise up the selector lever quadrant.
6 With the front seats fully to the rear, swivel the console and withdraw it at the same time reaching under the console and holding the plastic inner gaiter downward.
7 Refitting is a reversal of removal.

29 Facia panel – removal and refitting

1 Disconnect the battery.
2 Remove the instrument panel (Chapter 10) and the centre console (Section 28, this Chapter).

3 Remove the radio as described in Chapter 10, Section 31.
4 Remove the steering column upper shrouds and the fuse box cover.
5 Remove the glove compartment, after having extracted the securing screws which are located inside it.
6 Release the heater ducts from the facia outlets.
7 Extract the two screws from each air outlet at the ends of the facia. These outlets connect with the door glass demister ducts when the doors are closed. Pull the ducts and their connecting trunking down from behind the facia panel (photo).
8 Pull the fresh air grilles from the facia.
9 Reach up behind the facia and remove all the facia fixing screws including those at the side brackets and side panels and the scuttle bracket on the passenger side (photos).
10 Prise out the plugs at the ends of the windscreen demister grilles to expose the fixing screws. Extract the screws (photos).
11 With the help of an assistant, withdraw the facia until all the wiring plugs can be disconnected. Disconnect the illumination wires for the clock, glovebox, instrument and other illumination.
12 The centre air ducts will release from the heater as the facia is withdrawn.
13 Refitting is a reversal of removal.

29.7 Facia end air outlets

29.9A Facia mounting bracket (driver's side)

29.9B Facia mounting bracket
(passenger side)

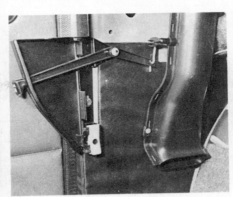

29.9C Side bracket on facia panel

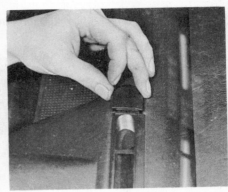

29.10A Windscreen demister grille screw plug

30 Sunroof panel – removal and refitting

1   Partially open the sunroof and release the clip which secures the front of the panel liner.
2   Open the trailing edge of the sunroof and pull the handle down. Press the button and turn the handle clockwise.
3   Ease both ends of the panel liner down and release the tensioner springs from their clips.
4   Close the sunroof.
5   Partially open the sunroof and slide the liner back between the panels.
6   Remove the panel liner tensioning springs.
7   Remove the sunroof panel followed by the weatherstrip and the trim tensioner clips.
8   To refit the sunroof panel, first locate it and then loosely secure it.
9   Close the panel and check its alignment.
10 Tighten the panel fixing screws and adjust the tilt arm as necessary.
11 Refit the tensioner springs.
12 Partially open the sunroof and then pull the liner forward making sure that the slides are under the spigots.
13 Open the trailing edge and engage the tensioning springs with the formed side of the clips. Close the sunroof then partially open it and secure the front liner clip.

31 Safety belts – care and maintenance

1   Periodically inspect the safety belts for fraying or other damage. If evident, renew the belt.
2   Cleaning of the belt fabric should be done with a damp cloth and a little detergent, nothing else.
3   Never alter the original belt anchorage and if the belts are ever removed, always take careful note of the sequence of mounting components. If the washers or collars are incorrectly positioned, the belt will not swivel as it has been designed to do.

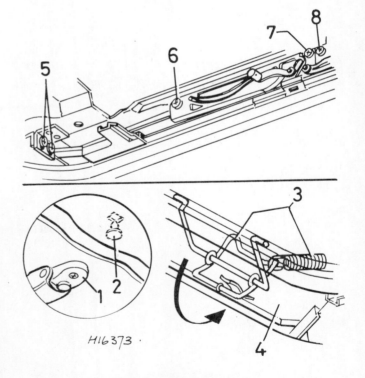

H16373

Fig. 12.22 Sunroof panel (Sec 30)

1 Control handle	5 Wind deflector slide
2 Front clip	screws
3 Spring tensioner and clip	6 Tilt arm front screw
4 Trim panel	7 Tilt arm rear screw
	8 Tilt arm adjuster screw

29.10B Windscreen demister grille screw

32 Rear view mirrors – removal and refitting

Interior
1   This is bonded directly to the windscreen glass.
2   Removal is usually carried out using a heat gun. When refitting, follow the instructions exactly as supplied with the pack of bonding agent.

Exterior
3   Open the front door and extract the two screws from the door glass demister duct (photo).
4   Support the mirror and remove the screws which secure the mirror adjuster retaining plate (photo).
5   Unscrew the adjuster knob and remove the mirror.
6   Refitting is a reversal of removal.

33 Grab handles

1   Grab handles are mounted on the roof lining above the door openings.
2   The securing screws are covered by a finisher which should be prised up using a small screwdriver inserted under its outer end to expose the screws (photo).

34 Parcels shelf – removal and refitting

1   Remove the two screws securing the ventilation control to the parcels tray, free the flap link and remove the control.
2   Remove the screws securing the shelf to the underside of the facia, and remove all other fixings to the heater and the facia support.
3   Remove the fresh air vents and withdraw the parcels shelf.
4   Refitting is a reversal of removal.

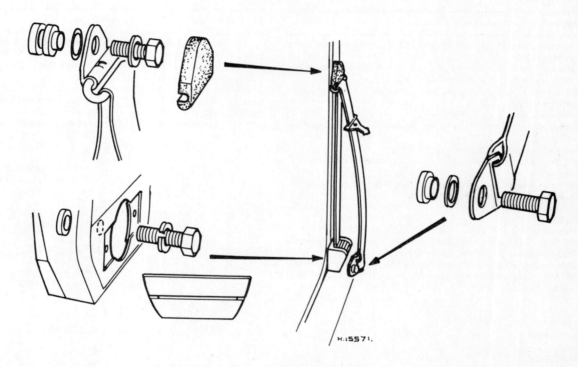

H.15571.

Fig. 12.23 Typical seat belt anchorages (Sec 31)

32.3 Door glass demister

32.4 Exterior mirror adjuster mounting plate

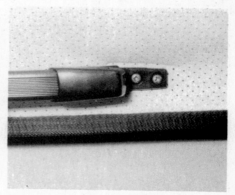

33.2 Grab handle mounting screw cover

Conversion factors

Length (distance)

	X		=		X		=	
Inches (in)	X	25.4	=	Millimetres (mm)	X	0.0394	=	Inches (in)
Feet (ft)	X	0.305	=	Metres (m)	X	3.281	=	Feet (ft)
Miles	X	1.609	=	Kilometres (km)	X	0.621	=	Miles

Volume (capacity)

	X		=		X		=	
Cubic inches (cu in; in³)	X	16.387	=	Cubic centimetres (cc; cm³)	X	0.061	=	Cubic inches (cu in; in³)
Imperial pints (Imp pt)	X	0.568	=	Litres (l)	X	1.76	=	Imperial pints (Imp pt)
Imperial quarts (Imp qt)	X	1.137	=	Litres (l)	X	0.88	=	Imperial quarts (Imp qt)
Imperial quarts (Imp qt)	X	1.201	=	US quarts (US qt)	X	0.833	=	Imperial quarts (Imp qt)
US quarts (US qt)	X	0.946	=	Litres (l)	X	1.057	=	US quarts (US qt)
Imperial gallons (Imp gal)	X	4.546	=	Litres (l)	X	0.22	=	Imperial gallons (Imp gal)
Imperial gallons (Imp gal)	X	1.201	=	US gallons (US gal)	X	0.833	=	Imperial gallons (Imp gal)
US gallons (US gal)	X	3.785	=	Litres (l)	X	0.264	=	US gallons (US gal)

Mass (weight)

	X		=		X		=	
Ounces (oz)	X	28.35	=	Grams (g)	X	0.035	=	Ounces (oz)
Pounds (lb)	X	0.454	=	Kilograms (kg)	X	2.205	=	Pounds (lb)

Force

	X		=		X		=	
Ounces-force (ozf; oz)	X	0.278	=	Newtons (N)	X	3.6	=	Ounces-force (ozf; oz)
Pounds-force (lbf; lb)	X	4.448	=	Newtons (N)	X	0.225	=	Pounds-force (lbf; lb)
Newtons (N)	X	0.1	=	Kilograms-force (kgf; kg)	X	9.81	=	Newtons (N)

Pressure

	X		=		X		=	
Pounds-force per square inch (psi; lbf/in²; lb/in²)	X	0.070	=	Kilograms-force per square centimetre (kgf/cm²; kg/cm²)	X	14.223	=	Pounds-force per square inch (psi; lbf/in²; lb/in²)
Pounds-force per square inch (psi; lbf/in²; lb/in²)	X	0.068	=	Atmospheres (atm)	X	14.696	=	Pounds-force per square inch (psi; lbf/in²; lb/in²)
Pounds-force per square inch (psi; lbf/in²; lb/in²)	X	0.069	=	Bars	X	14.5	=	Pounds-force per square inch (psi; lbf/in²; lb/in²)
Pounds-force per square inch (psi; lbf/in²; lb/in²)	X	6.895	=	Kilopascals (kPa)	X	0.145	=	Pounds-force per square inch (psi; lbf/in²; lb/in²)
Kilopascals (kPa)	X	0.01	=	Kilograms-force per square centimetre (kgf/cm²; kg/cm²)	X	98.1	=	Kilopascals (kPa)
Millibar (mbar)	X	100	=	Pascals (Pa)	X	0.01	=	Millibar (mbar)
Millibar (mbar)	X	0.0145	=	Pounds-force per square inch (psi; lbf/in²; lb/in²)	X	68.947	=	Millibar (mbar)
Millibar (mbar)	X	0.75	=	Millimetres of mercury (mmHg)	X	1.333	=	Millibar (mbar)
Millibar (mbar)	X	0.401	=	Inches of water (inH₂O)	X	2.491	=	Millibar (mbar)
Millimetres of mercury (mmHg)	X	0.535	=	Inches of water (inH₂O)	X	1.868	=	Millimetres of mercury (mmHg)
Inches of water (inH₂O)	X	0.036	=	Pounds-force per square inch (psi; lbf/in²; lb/in²)	X	27.68	=	Inches of water (inH₂O)

Torque (moment of force)

	X		=		X		=	
Pounds-force inches (lbf in; lb in)	X	1.152	=	Kilograms-force centimetre (kgf cm; kg cm)	X	0.868	=	Pounds-force inches (lbf in; lb in)
Pounds-force inches (lbf in; lb in)	X	0.113	=	Newton metres (Nm)	X	8.85	=	Pounds-force inches (lbf in; lb in)
Pounds-force inches (lbf in; lb in)	X	0.083	=	Pounds-force feet (lbf ft; lb ft)	X	12	=	Pounds-force inches (lbf in; lb in)
Pounds-force feet (lbf ft; lb ft)	X	0.138	=	Kilograms-force metres (kgf m; kg m)	X	7.233	=	Pounds-force feet (lbf ft; lb ft)
Pounds-force feet (lbf ft; lb ft)	X	1.356	=	Newton metres (Nm)	X	0.738	=	Pounds-force feet (lbf ft; lb ft)
Newton metres (Nm)	X	0.102	=	Kilograms-force metres (kgf m; kg m)	X	9.804	=	Newton metres (Nm)

Power

	X		=		X		=	
Horsepower (hp)	X	745.7	=	Watts (W)	X	0.0013	=	Horsepower (hp)

Velocity (speed)

	X		=		X		=	
Miles per hour (miles/hr; mph)	X	1.609	=	Kilometres per hour (km/hr; kph)	X	0.621	=	Miles per hour (miles/hr; mph)

Fuel consumption*

	X		=		X		=	
Miles per gallon, Imperial (mpg)	X	0.354	=	Kilometres per litre (km/l)	X	2.825	=	Miles per gallon, Imperial (mpg)
Miles per gallon, US (mpg)	X	0.425	=	Kilometres per litre (km/l)	X	2.352	=	Miles per gallon, US (mpg)

Temperature

Degrees Fahrenheit = (°C x 1.8) + 32

Degrees Celsius (Degrees Centigrade; °C) = (°F - 32) x 0.56

*It is common practice to convert from miles per gallon (mpg) to litres/100 kilometres (l/100km), where mpg (Imperial) x l/100 km = 282 and mpg (US) x l/100 km = 235

Index

Printed by
J H Haynes & Co Ltd
Sparkford Nr Yeovil
Somerset BA22 7JJ England